Lecture Notes in Biomathematics

Managing Editor: S. Levin

43

Mathematical Modeling of the Hearing Process

Proceedings, Troy, NY 1980

Edited by Mark H. Holmes and Lester A. Rubenfeld

Springer-Verlag
Berlin Heidelberg New York

Lecture Notes in Biomathematics

Vol. 1: P. Waltman, Deterministic Threshold Models in the Theory of Epidemics. V, 101 pages. 1974.

Vol. 2: Mathematical Problems in Biology, Victoria Conference 1973. Edited by P. van den Driessche. VI, 280 pages. 1974.

Vol. 3: D. Ludwig, Stochastic Population Theories. VI, 108 pages. 1974.

Vol. 4: Physics and Mathematics of the Nervous System. Edited by M. Conrad, W. Güttinger, and M. Dal Cin. XI, 584 pages. 1974.

Vol. 5: Mathematical Analysis of Decision Problems in Ecology. Proceedings 1973. Edited by A. Charnes and W. R. Lynn. VIII, 421 pages. 1975.

Vol. 6: H. T. Banks, Modeling and Control in the Biomedical Sciences. V, 114 pages. 1975.

Vol. 7: M. C. Mackey, Ion Transport through Biological Membranes, An Integrated Theoretical Approach. IX, 240 pages. 1975.

Vol. 8: C. DeLisi, Antigen Antibody Interactions. IV, 142 pages. 1976.

Vol. 9: N. Dubin, A Stochastic Model for Immunological Feedback in Carcinogenesis: Analysis and Approximations. XIII, 163 pages. 1976.

Vol. 10: J. J. Tyson, The Belousov-Zhabotinskii Reaktion. IX, 128 pages. 1976.

Vol. 11: Mathematical Models in Medicine. Workshop 1976. Edited by J. Berger, W. Bühler, R. Repges, and P. Tautu. XII, 281 pages. 1976.

Vol. 12: A. V. Holden, Models of the Stochastic Activity of Neurones. VII, 368 pages. 1976.

Vol. 13: Mathematical Models in Biological Discovery. Edited by D. L. Solomon and C. Walter. VI, 240 pages. 1977.

Vol. 14: L. M. Ricciardi, Diffusion Processes and Related Topics in Biology. VI, 200 pages. 1977.

Vol. 15: Th. Nagylaki, Selection in One- and Two-Locus Systems. VIII, 208 pages. 1977.

Vol. 16: G. Sampath, S. K. Srinivasan, Stochastic Models for Spike Trains of Single Neurons. VIII, 188 pages. 1977.

Vol. 17: T. Maruyama, Stochastic Problems in Population Genetics. VIII, 245 pages. 1977.

Vol. 18: Mathematics and the Life Sciences. Proceedings 1975. Edited by D. E. Matthews. VII, 385 pages. 1977.

Vol. 19: Measuring Selection in Natural Populations. Edited by F. B. Christiansen and T. M. Fenchel. XXXI, 564 pages. 1977.

Vol. 20: J. M. Cushing, Integrodifferential Equations and Delay Models in Population Dynamics. VI, 196 pages. 1977.

Vol. 21: Theoretical Approaches to Complex Systems. Proceedings 1977. Edited by R. Heim and G. Palm. VI, 244 pages. 1978.

Vol. 22: F. M. Scudo and J. R. Ziegler, The Golden Age of Theoretical Ecology: 1923–1940. XII, 490 pages. 1978.

Vol. 23: Geometrical Probability and Biological Structures: Buffon's 200th Anniversary. Proceedings 1977. Edited by R. E. Miles and J. Serra. XII, 338 pages. 1978.

Vol. 24: F. L. Bookstein, The Measurement of Biological Shape and Shape Change. VIII, 191 pages. 1978.

Vol. 25: P. Yodzis, Competition for Space and the Structure of Ecological Communities. VI, 191 pages. 1978.

Lecture Notes in
Biomathematics

Managing Editor: S. Levin

43

T.M.

Mathematical Modeling of the Hearing Process

Proceedings of the NSF-CBMS Regional Conference
Held in Troy, NY
July 21–25, 1980

Edited by Mark H. Holmes and Lester A. Rubenfeld

Springer-Verlag
Berlin Heidelberg New York 1981

Editors

Mark H. Holmes
Lester A. Rubenfeld
Department of Mathematical Sciences
Rensselaer Polytechnic Institute
Troy, NY 12181, USA

AMS Subject Classifications (1980): 73P05, 76Z05, 92-02

ISBN 3-540-11155-7 Springer-Verlag Berlin Heidelberg New York
ISBN 0-387-11155-7 Springer-Verlag New York Heidelberg Berlin

Printing and binding: Beltz Offsetdruck, Hemsbach/Bergstr.
2141/3140-543210

PREFACE

The articles of these proceedings arise from a NSF-CBMS regional conference on the mathematical modeling of the hearing process, that was held at Rensselaer Polytechnic Institute in the summer of 1980. To put the articles in perspective, it is best to briefly review the history of such modeling. It has proceeded, more or less, in three stages. The first was initiated by Herman Helmholtz in the 1880's, whose theories dominated the subject for years. However, because of his lack of accurate experimental data and his heuristic arguments it became apparent that his models needed revision. Accordingly, based on the experimental observations of von Bekesy, the "long wave" theories were developed in the 1950's by investigators such as Zwislocki, Peterson, and Bogert. However, as the experiments became more refined (such as Rhode's Mossbauer measurements) even these models came into question. This has brought on a flurry of activity in recent years into how to extend the models to account for these more recent experimental observations. One approach is through a device commonly refered to as a second filter (see Allen's article) and another is through a more elaborate hydroelastic model (see Chadwick's article). In conjunction with this latter approach, there has been some recent work on developing a low frequency model of the cochlea (see Holmes' article). Over the last few years there has arisen a discrepancy between even different experiments, in particular, between mechanical or structural measurements (such as in Rhode's work) and neural measurements (as in Kiang's work). The article by Khanna and Leonard addresses this fundamental question. Indeed, at the risk of being presumptive, the resolution of this problem should lead to the fourth, and possibly definitive, stage of the modeling.

The articles mentioned so far deal exclusively with modeling the dynamics of the cochlea. Another major component of the hearing apparatus is the semi-circular canals. An overview of the modeling of this interesting organ can be found in Van Buskirk's article. Along different lines, the article by Sondhi presents some of his interesting work on inverse problems that arise from models of the cochlea.

We would like to thank the National Science Foundation and the Conference Board of Mathematical Sciences for their support of this conference.

Mark H. Holmes
Lester A. Rubenfeld

TABLE OF CONTENTS

Cochlear Modeling-1980 by J. B. Allen 1

Studies in Cochlear Mechanics by R. S. Chadwick 9

A Hydroelastic Model of the Cochlea: An Analysis for Low Frequencies by M. H. Holmes 55

Basilar Membrane Response Measured in Damaged Cochleas of Cats by S. M. Khanna and D. G. B. Leonard 70

A Mathematical Model of the Semicircular Canals by W. C. Van Buskirk 85

The Acoustical Inverse Problem for the Cochlea by M. M. Sondhi 95

Cochlear Modeling - 1980

J. B. Allen
Acoustics Research Dept.

Bell Laboratories
Murray Hill, New Jersey 07974

ABSTRACT

A general review of the present state of cochlear mechanical modeling is presented in this paper, and some new experimental neural phase data is discussed and compared to a recent theory of hair cell transduction. This theory of transduction unifies experimental mechanical and neural tuning data while it simultaneously predicts a 180° phase shift in the neural response one octave below CF. The new phase measurements qualitatively corroborate this transduction theory. The intent of this paper is to be somewhat philosophical about the future directions of cochlear modeling and cochlear theory and it attempts to provide an alternative explanation of many widely accepted diverse experimental results.

Introduction

In this paper I would like to review and discuss some theoretical and experimental aspects of modern hearing theory. In many ways this theory is still quite sketchy and controversial. Therefore it is difficult to present a point of view which might remain unchallenged. On the other hand, many pieces of the complex experimental puzzle seem to fit together. In this paper I am going to present that experimental puzzle. I will also discuss a theory that explains many divergent pieces of otherwise unrelated data.

To understand the hearing process is to understand the cochlea, since it is that organ which transforms mechanical energy, in the form of sound waves, into the neural code, which is then interpreted by the brain as sound. It is now recognized that the basic properties of monaural sound perception are established in the cochlea: critical bands, masking, loudness, pitch and frequency discrimination. The study of hearing over the last hundred years has therefore been focused on understanding and modeling the function of the cochlea.

The anatomy of the cochlea is quite amazing. It is an electro-mechanical organ machined by nature with a precision that exceeds by orders of magnitude the resolution of our present integrated circuit technology. In Figure 1 we show a cross section through the cochlea, which is planted deep in the temporal bone. It is the cochlea's inaccessibility that makes it so difficult to study, unlike the vocal cords or the eye which are relatively easily probed optically. Sound enters the ear canal via the pinna folds and vibrates the ear drum, or tympanic membrane. The tympanic membrane is connected to a three bone lever system (the ossicles). One of these small bones, the stapes, drives the fluid in the scala vestibuli.

Anatomy of the Cochlea

It is widely believed that the cochleae of all mammals function in a similar manner (e.g. the physical principles are the same) since the cochleae of humans, bats, cats, etc. are similar in their anatomy their neural response. In order to understand human hearing therefore, it is useful to study the mammalian cochlea.

If the cochlea could be uncoiled, its cross section would consist of three chambers as shown in Fig. 1. The central partition is called the cochlear partition and it includes all of the chamber called the scala media (SM). This chamber is ionically distinct from the ionic composition of the scala vestibuli (SV) or scala tympani (ST). The scala media is also electrically isolated from the other two chambers. On the side wall of SM is the stria vascularis, a very vascular area having a large potential drop across it. The stria functions as a battery with a driving current which is used by the hair cells in the mechanical to electrical transduction process. Reissner's membrane separates scala media from scala vestibuli, and acts to isolate the fluids of the spaces both electrically and chemically. It is believed that Reissner's membrane is unimportant from a mechanical point of view [Steel (1974)].

The partition separating the scale media from the scala tympani is the basilar membrane. The basilar membrane (BM) is mechanically tuned with a spatially varying resonant frequency. It is very stiff near the stapes, and becomes progressively less stiff along its length. In cats, the resonant frequency varies from 40 kHz at the stapes to about 100 Hz at the helicotrema, the far end. The resonant frequency of the BM results from a spring-mass resonance for which the resonant frequency f_r at each point x is given by

$$f_r(x) = \frac{1}{2\pi} \sqrt{K(x)/m} \qquad (1)$$

where $K(x)$ is the basilar membrane spring stiffness and m is the BM mass.

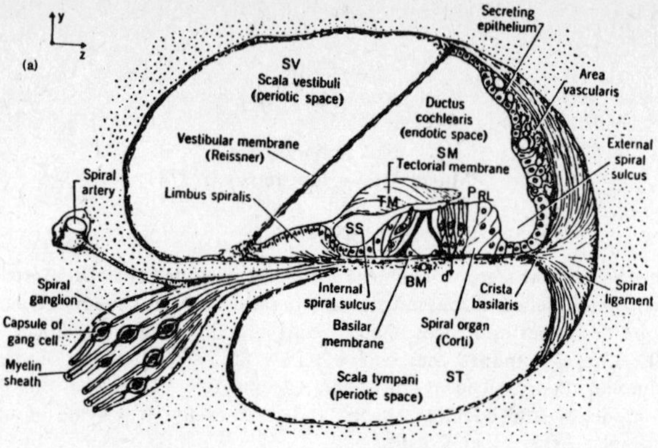

Fig. 1 Drawing of a cross section of the cochlear canal due to Rassmussen. p is the pressure in scala media while d represents the displacement of the basilar membrane.

According to calculations, it appears that the mechanical coupling (in the form of a bending moment) along the length of the basilar membrane is mainly through the fluid medium (coupling does not take place through the basilar membrane proper). According to this principle, the basilar membrane is a plate (or bar) along the width axis, but acts like a membrane longitudinally. Such a structure would spring back if bent along the width axis, but would display no restoring forces if bent longitudinally.

Thus the basilar membrane is modeled as a distributed second order (spring-mass) resonance system driven by the surrounding fluid medium. However because of the fluid coupling between the sections, the actual frequency response at each point is not that of a simple second order resonance. Instead the point frequency response is strongly modified. In fact the cochlea acts much like a dispersive delay line having a spatially varying cutoff frequency.

A simple qualitative model might clarify this point. Suppose we were to cascade a large number of low-pass filters together in such a way that each successive low-pass filter had a slightly lower cutoff frequency. If one were to excite this transmission line with a low frequency tone burst, the burst would pass through each of the low pass filters unattenuated, but delayed. As the tone burst reached the filter having its cutoff at the frequency of the tone burst, the signal would be attenuated. Such a frequency dependent transmission line would have much in common with the cochlea. This model is similar to a model proposed by Kim, Molnar and Pfeiffer (1973). The dispersive line effect is what was observed by Von Bekesy in his early experiments. The first quantification of the dispersive delay line model was due to J. J. Zwislocki (1948). Today of course, because of digital computers, we have been able to greatly refine these early models by numerical methods [Allen & Sondhi (1979)].

Cochlear Macromechanics

Cochlear theory may be divided into two parts: cochlear macromechanics and micromechanics. The first division includes the fluid, and treats the basilar membrane as a point impedance $Z(x,\omega) = -2p(x,\omega)/u(x,\omega)$, where $2p(x,\omega)$ is the pressure across the basilar membrane, and $u(x,\omega)$ is its velocity. Excitation is due to the stapes and round window which are modeled as pistons moving in opposite phase. Since the cochlear fluid is incompressible, it may be described by Laplace's equation in the fluid pressure, independent of time or frequency

$$\nabla^2 p(x,y)=0 .$$

This equation results from assuming that the velocity of sound is great enough that a wavelength is larger than the dimensions of the cochlea. Since the fluid is incompressible, the volume integral of BM displacement must be equal to the total fluid volume displaced at the stapes. For a constant frequency stimulus, the BM looks as if it has a hole in it at any point where the BM impedance is zero, namely where the stiffness reactance exactly cancels the mass reactance. Thus most of the motion takes place near this resonant point. In fact it takes place before the resonant point (on the stapes side of resonance).

Finding the solution of the macromechanical model requires the solution of Laplace's equation in a box having one wall (the basilar membrane) specified by the spatially varying impedance. The most common form for this impedance is

$$Z(x,\omega)=\frac{K(x)}{i\omega}+R(x)+i\omega m \tag{2}$$

where $K(x)$ specifies the basilar membrane stiffness along its length, $R(x)$ specifies the acoustic resistance, and m is the mass. This problem may be solved exactly using Green's functions, by which one may transform the problem into an integral equation (other exact solution methods have also been quite successful, for example Neely (1977), Viergever (1980)). The integral equation may be expressed as a circular convolution which may be evaluated by FFT methods. These details are beyond the scope of this review but they may be found in Allen and Sondhi (1979). The important ideas we wish to emphasize here are the concepts of an incompressible fluid and the point impedance model of the basilar membrane.

From exact numerical solutions of the two dimensional incompressible fluid model we find an interesting picture. The macromechanical system acts as a delay line which has a very strong frequency dependence. In the cochlea, the signal energy is propagated with a delay and cutoff that is position dependent. As a result of the position (place) dependent delay, energy is progressively removed as the signal travels along the basilar membrane, the higher frequencies first and the lower frequencies as the signal reaches the far end of the cochlea.

In order to show the delay line effect analytically it is necessary to make several simplifying assumptions. First it is possible to derive an approximate low frequency theory in the form of a second order transmission line model [Sondhi (1978)]. This theory only gives approximate results at the cutoff frequency, but it is a very accurate theory for frequencies below cutoff [Allen (1978)]. The transmission line equation may then be evaluated by an approximate solution method called the WKB method [Zwieg, Lipes, Pierce (1975)]. Very little error is introduced by the WKB solution method.

Figure 2 shows the steady state pure tone velocity envelope frequency response at various points on the basilar membrane as found from the full two-dimensional model. Also shown is an experimental result due to Rhode (1971) as measured in squirrel monkey. The response rises with a slope of 6 to 9 db/oct which then increases to about 24 dB/oct at the maximum. Finally, the response cuts off very sharply with a cutoff slope that can be greater than 500 dB/oct. In bats this slope reaches 1000 dB/oct. The frequency response of each point on the BM is similar in its shape, but the frequency of its cutoff varies with position. The relation

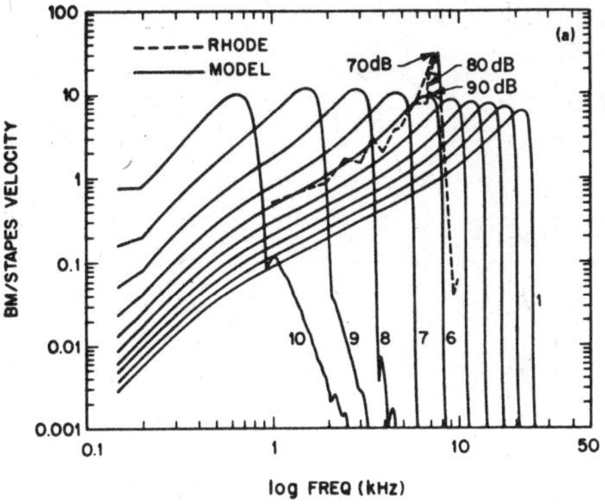

Fig. 2 Model calculations made from the two dimensional model of Allen & Sondhi (1979) for ten stations along the basilar membrane. The dashed lines are the data of Rhode (1971). Note the nonlinear effects near the best frequency.

between the cutoff frequency and position along the basilar membrane is called the cochlear map. The highest frequencies appear near the stapes and the cutoff frequency decreases with distance from the stapes. The ten responses in Fig. 2 are for equally spaced stations along the basilar membrane.

Cochlear Nonlinearities

A very significant feature of Rhode's data was that he found a compressive nonlinearity at frequencies neighboring the cutoff frequency (see Fig. 2). As a result, the output (BM displacement or velocity) varies much less than the stapes input displacement or velocity for frequencies near the best frequency. The significance of this important finding will become clearer as we proceed, but, in my opinion, it is a precursor to an automatic gain control system which seems to be built into the cochlear filters. Note that even though the cochlea is nonlinear, it still operates as a low-pass filter having a very sharply defined cutoff. The nonlinearity generates distortion products, as would be expected of any nonlinear system. Particularly important are intermodulation distortion products having frequencies which are lower than the primary frequencies since these distortions are further propagated along the cochlea to their place [Kim (1980) Hall (1980)] (recall the delay line description of the macromechanics). Harmonic distortion however is generated at a place beyond its cutoff since it is always greater in frequency than the fundamental. Since it is generated in the cutoff region beyond its resonant place, the distortion signal cannot propagate back to its place (its cutoff point) because the local propagation constant for the distortion frequency is real. This explains the ear's greater sensitivity to sub-harmonic distortion. The automatic gain control nonlinearity also explains why the harmonic distortion is always below the primarys in intensity and does not grow large at large input levels as would be predicted from a power law nonlinearity. It is presently believed in fact that the intermodulation distortion never seems to be greater than -15 dB equivalent ear canal sound pressure level relative to the primary signals.

The source of the nonlinearity remains unknown, although there are some good reasons to believe that its generation is related to motions of the stereocilia of the outer hair cells. It presently seems clear that the source of distortion is not the byproduct of some poorly engineered cochlear component. The distortion is rather perhaps the negligible residual of a sophisticated local feedback mechanism in the mechanical motion of the properly operating cochlea, such as the automatic gain control system mentioned previously.

Cochlear Micromechanics

The second major area of cochlear theory, besides macromechanics, is micromechanics, which deals with the physics of the organ of Corti. Figure 3 shows a cross section through the organ of Corti, which includes the basilar membrane, the tectorial membrane, and the inner and outer hair cells. The tectorial membrane is clearly involved in the acoustic signal path. The most commonly accepted model assumes that when the BM moves, the tectorial membrane and the reticular lamina shear. This type of motion is similar to the motion of a pantograph. Figure 4 shows how this shearing is believed to take place. In this model, as in the cochlea, the TM and RL are connected through the stereocilia of the outer hair cells. These hairs are most likely stiff to the shear displacements, due to their actin cross-linked composition [Tilney, et al. (1980)]. Any shearing stiffness in the stereocilia would, according to this model, add stiffness to the bending stiffness of the basilar membrane.

From Figure 3 we see two groups of hair cells - one row of inner hair cells, and three rows of outer hair cells. It is only the inner hair cells that we "hear" with, since almost all afferent nerve fibers (neurons which carry signals to the brain) start at the synapses located at the base of these gourd shaped cells. The outer hair cells on the other hand are part of the efferent system (the neural system which carries signals from the brain). Clearly the 64 dollar question is: "What is the purpose of the outer hair cells?" We shall come back to this question.

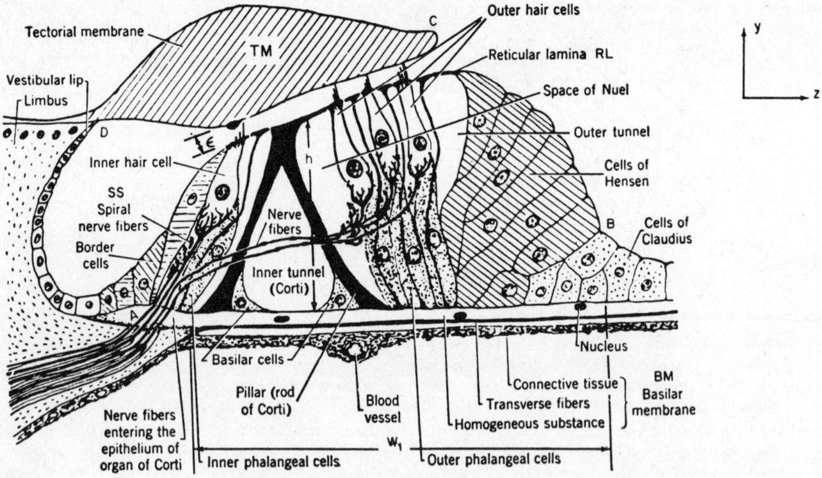

Fig. 3 A detailed view of the organ of Corti.

We have seen the tuning response of the basilar membrane; we now look at the tuning properties of neurons and hair cells. In the last few years one laboratory has been successful in recording the voltage inside inner hair cells. The experiment was very difficult, needless to say, but what they found made their effort totally worthwhile [Russell & Sellick (1978)]. They found the tuning to be much sharper than that of the basilar membrane and that the nonlinearity (level effects) seemed to be much greater. When these curves were compared to neurally measured tuning curves, they had quite similar tuning. In Figs. 5a and 7 we show typical neural tuning curves. These curves give the sound pressure level necessary to increase the spike rate on the neuron as a function of frequency.

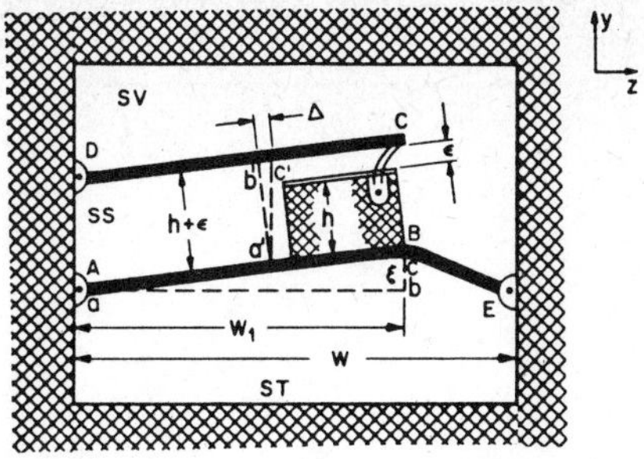

Fig. 4 A cross-section showing the BM displaced. Since AD and BC are equal in length in this simplified model, AB remains parallel to DC after any BM displacement ξ. As a result, triangles abc and a'b'c' are similar, and displacement Δ is proportional to the BM displacement ξ.

Is the difference in mechanical and neural tuning real, or is it an artifact of the measurement method? We now know that the nonlinear behavior is very sensitive to the slightest cochlear damage. Would more carefully executed basilar membrane measurements (under conditions of less damage) show a tuning which is similar to the hair cell receptor potential or neural tuning?

While this possibility can certainly not be ruled out, and in fact many investigators believe it to be likely, a second group feel that a true difference exists between BM response and hair cell or neural response. Compelling arguments lie on both sides of this issue.

Thus it appears that without a much better understanding of the transduction mechanics (e.g., better cochlear micromechanical models), and of the nonlinearity of the basilar membrane, we will probably not be able to understand intensity coding, or even frequency discrimination, at the neural level.

The "Second-Filter" Hypothesis

In order to explain the observed differences between BM tuning and neural tuning, the concept of a "second-filter" was introduced. Originally this filter, in fact, was not really a filter at all since it included all non-linear effects that were not understood at the time. Today the concept of a second filter is still reasonable, but in a much simpler form.

If one adds to the macromechanical model the concept of a *transduction filter* $H_T(x,s)$ consisting of a second order pole - zero pair, namely

$$H_T(x,s) = \frac{s^2 + 2\zeta_z(x)\, s + s_z^2(x)}{s^2 + 2\zeta_p(x)\, s + s_p^2(x)} \tag{3}$$

where ζ_p, ζ_z, are damping parameters of the model, $s = i\omega$, and $\omega_p(x)$ and $\omega_z(x)$ are position dependent frequency parameters, then it is possible to bridge the gap between model macromechanical BM response and neural or receptor potential response. Not only is it possible to calculate neural tuning curves in this way, but certain very interesting predictions about neural phase result.

In order that $H_T(x,s)$ properly sharpen the low pass BM frequency response, the zeros of H_T must be one octave below the cutoff frequency [Allen, (1980)]. At that frequency we would therefore expect a phase shift consistent with a zero (or anti-resonance). According to the model (and the best experimental data to date) this phase shift would be present in the neural signal but not present in the BM response.

Neural Phase

In an earlier paper [Allen, (1980)], the need for the transduction filter of Eq. 3 is discussed in some detail, and a prediction is made about the neural phase response in the frequency region near the zero frequency ω_z. From Eq. 3, the neural phase should undergo a π phase shift. According to earlier estimates, the zero should be about one octave below the characteristic frequency CF. We now present neural phase data that demonstrate that this is the case. In Fig. 5 we show the threshold tuning curve and phase data for a cat neural unit. The threshold tuning was measured by increasing a pure tone level at each frequency until an increase in the neural spike rate was detected. The phase (Fig. 5b) was measured by sweeping a pure tone through a super threshold range of the tuning curve (as shown by the solid lines of Fig. 5a) and forming a stimulus-locked spike histogram (PST). Phase was calculated from the phase of the Fourier transform of the PST at the stimulation frequency. Since a pure delay corresponds to a linearly decreasing phase, a linear "regression" line having a negative slope corresponding to a fixed delay was subtracted from the phase data in order to remove acoustic and cochlear delays and keep the phase variations within $\pm\pi$ radians. In this way, it was not necessary to correct for 2π phase jumps which would otherwise occur. In Figure 5b, 2.7 msec of delay was removed.

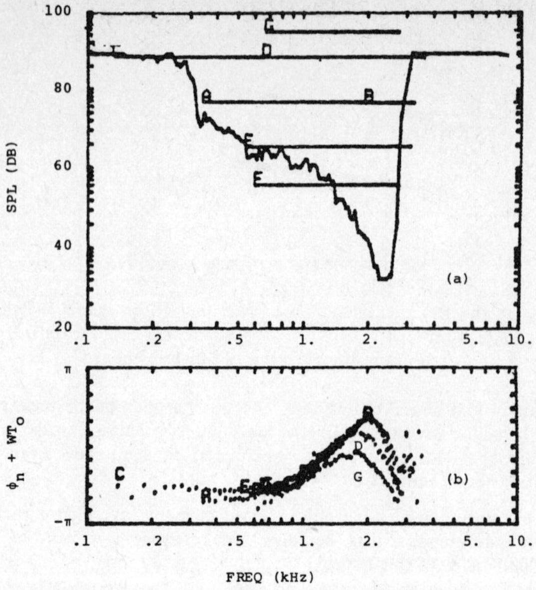

Fig. 5 (a) Neural tuning data for a cat fiber. The curve shows the sound pressure level required to cause a 20% increase in neural spike rate above the spontaneous rate due to a 50 ms tone burst. (b) Phase of the neuron corrected for $T_0 = 2.7$ ms of acoustical and cochlear delay. Note the nonlinear phase effects for 90 and 95 dB SPL (sweeps D and G of (a)).

This delay would correspond to a phase curve having negative slope and passing through $\phi = 0$ at $\omega = 0$. The true neural phase ϕ_n is therefore related to ϕ, the phase of Fig. 5b, by the relation

$$\phi_n(\omega) = \phi(\omega) - \omega\, T_0 \tag{4}$$

where $T_0 = 2.7$ msec.

The interesting feature in the data is the region of positive phase slope one octave below *CF*. As discussed, this phase jump is consistent with the spectral zero model of transduction. No other models presently give a π phase shift. The frequency limits of the π phase shift for a large number of units in one cat is shown in Fig. 6 plotted against each neurons best frequency f_{CF}. Note that the zero, as measured by the phase shift, occurs one octave below *CF*.

Two Tone Suppression

We now briefly present new measurements on cat neurons made in the presence of a sub-threshold second tone. In Fig. 7 we see a family of five tuning curves. The lowest threshold response was taken with no second tone present, while the four other responses were measured in the presence of second tones having magnitudes and frequencies as shown by the letters A,B,C,D. Apparently the effect of the sub-threshold second tone is to modify the frequency response of the excitatory neural signal. Certain properties of this phenomenon, which is called two tone suppression, have been carefully studied.

Two tone suppression is defined as the locus of levels and frequencies of a suppressor tone which will suppress a fixed super-threshold CF tone having a level which is 10 dB above threshold [Sachs & Kiang (1969)]. From Fig. 7, we see that a CF tone 10 dB above the unsuppressed threshold would be below threshold in the presents of tones A, B, or C since those three suppressors all give rise to at least 10 dB of threshold elevation at CF. As defined, "two-tone suppression" is a threshold effect because of what that paradigm measures. The data of Fig. 7 on the other hand show the super threshold effects of this cochlear nonlinearity. In fact, sub-threshold suppressor tone A suppresses a CF tone by 40 dB. Clearly the effect of the cochlear nonlinearity is functionally significant.

An open question is what is the source of this nonlinearity effect? It is related to the nonlinear basilar membrane response as seen by Rhode Fig. 2, and is, I believe, characteristic of an automatic gain control. Understanding and modeling this nonlinear effect is at the forefront of modern cochlear research [Kim (1980); Hall (1980)].

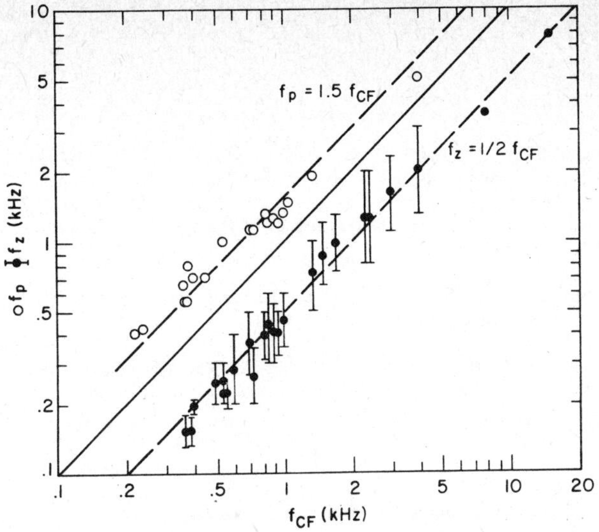

Fig. 6 This figure shows the frequency regions of positive going phase for 23 different units as a function of CF for one cat. The solid circles mark the geometric mean of the phase shift region. The open circles mark frequencies where "bumps" occur on high frequency side of the best frequency as a function of the best frequency. These "bumps" do not appear in either Figs. 5a or 7.

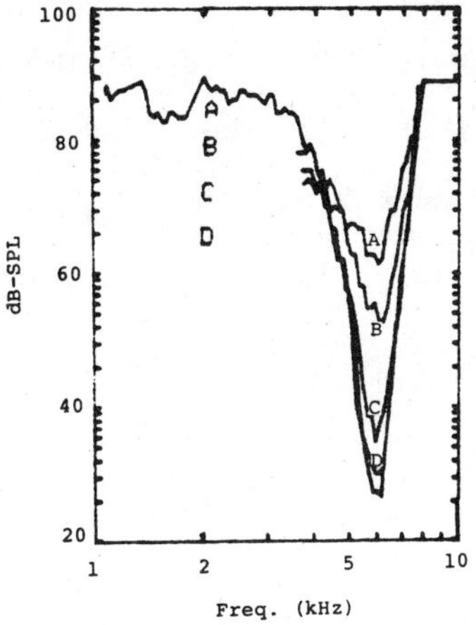

Fig. 7 Tuning curves made in the presence of a second tone having a frequency and level as shown by the letters A,B,C,D. The threshold is greatly altered (by up to 40 dB) by the sub-threshold second tone. No linear system would show such level dependent frequency behavior.

References

Allen, J. B., (1980). "Cochlear micromechanics -- A physical model of transduction," J. Acoust. Soc. Am. *68*, 1660-1670.

Allen, J. B., and Sondhi, M. M., (1979). "Cochlear macromechanics: Time domain solutions," J. Acoust. Soc. Am. *66*, 123-132.

Allen, J. B., (1979). "Cochlear Models - 1978," in "Models of the Auditory System and Related Signal Processing Techniques," Eds: M. Hoke and E. de Boer, *Scan of Audiol. Suppl. 9*, pp. 1-16.

Hall, J. L., (1980). "Cochlear Models: Evidence in Support of Mechanical Nonlinearities and a Second Filter (A Review)," Hearing Research *2*, 455-464.

Kim, D. O., (1980). "Cochlear Mechanics: Implications of Electrophysiological and Acoustical Observation," Hearing Research *2*, 297-317.

Kim, D. O., Molnar C. E., Pfeiffer, R. R., (1973). "A System of Nonlinear Differential Equations Modeling Basilar-Membrane Motion," J. Acoust. Soc. Am. *54*, 1517-1529.

Neely, S. T., (1977). "Mathematical Models of the Mechanics of the Cochlea," California Inst. of Tech. Eng. Thesis, Thesis Advisor, J. R. Pierce.

Rhode, W. S. (1971). "Observations of the Vibration of the Basilar Membrane in Squirrel Monkeys using the Mossbauer Technique," J. Acoust. Soc. Am. *49*, 1218-1231.

Russell, I. J., and Sellick, P. M., (1978). "Intracellular Studies of Hair Cells in the Mammalian Cochlea," J. Physiol. *284*, 261-290.

Sachs, M. B. & Kiang, Y. S. (1969). "Two-tone Inhibition in Auditory-Nerve Fibers," J. Acoust. Soc. Am., *43*, 1120-1968.

Sondhi, M. M. (1978). "A Method for Computing Motion in a Two-Dimensional Model," J. Acoust. Soc. Am. *63*, 1468-1477.

Steel, C. R. (1974). "Stiffness of Reissner's Membrane," J. Acoust. Soc. Am., *56*, 1252-1257.

Tilney, L. G., De Rosier, D. J., and Mulroy, M. J., (1980). "The Organization of Actin Filaments in the Stereocilia of Cochlear Hair Cells," J. Cell Biol. *85*, 244-259.

Viergever, M. A. (1980). *Mechanics of the inner ear*, PhD. Thesis, Delft University Press.

Zwieg, G., Lipes, R., and Pierce, J. R., (1976). "The Cochlear Compromise," J. Acoust. Soc. Am. *59*, 975-982.

Zwislocki, J. (1948). Acta OtoLaryngol, Suppl. 72 (in German). See also Zwislocki, J. (1950), "Theory of Acoustical Action of the Cochlea," J. Acoust. Soc. Am. *22*, 778-784.

Studies in Cochlear Mechanics

R.S. Chadwick

Biomedical Engineering and Instrumentation Branch

Division of Research Services

National Institutes of Health

Bethesda, MD 20205

I. Introduction and Scope

These notes represent some of the efforts by the author to understand mechanisms which have relevance to the mechanics of the cochlea. The use of asymptotic methods is stressed. In particular, the "WKB" method or its generalization, the two-variable expansion procedure (Cole, 1968, Chapter 3) is used throughout. The notes are arranged in the format of three specific problems, in the order of increasing relevance and difficulty. Section II deals with the sinusoidal forced vibrations of an isotropic membrane with a tapered planform in the absence of a fluid. Section III considers the response of a partition having mass, internal damping and variable stiffness and which is immersed in an inviscid fluid undergoing two dimensional motions. Section IV illustrates some aspects of the response to sinusoidal forcing of a highly anisotropic tapered plate immersed in a viscous fluid undergoing three dimensional motions. Special consideration is given to the low and high frequency limits of this problem.

These problems do not deal with potentially important mechanical nonlinearities, or even linear "micro-mechanical" effects, i.e. motions associated with the detailed structural elements of the Organ of Corti. As such, we are concerned here with the gross response or "macro-mechanics" of the basilar membrane.

II. Forced Vibration of a Tapered Membrane

A convenient starting point is the problem of the forced response of a long tapered membrane vibrating in the absence of fluid. This simplified system has many features in common with the hydro-elastic problem of the cochlea, but the calculations are much simpler. The forced response of a slightly tapered, nearly rectangular membrane was previously considered by Goodier and Fuller (1964). They found that for the same near resonant excitation of a linearly tapered membrane the response is radically different depending on which end is forced. For excitation at the wider end, there can be a large amplitude response in the wider half, while the narrower half remains almost undisturbed. For excitation at the narrower end there is only a small amplitude response throughout. They attribute this peculiarity to the presence of a turning point. We reconsider this problem here in a slightly more general form and focus on the response to narrow end excitation, since this more closely resembles the response in the cochlea. Our approach utilizes fast and slow longitudinal space coordinates to cope with the slowly varying geometry, and a transition layer which connects oscillatory solutions to exponential solutions when needed. Chadwick (1978) used the same approach to study the free vibrations of long narrow tapered plates.

The geometry of the membrane planform is shown in Figure 1. In terms of the scaled

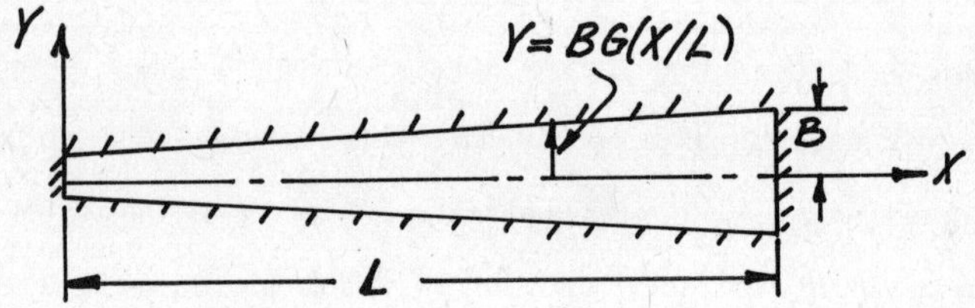

Figure 1. Geometry of membrane planform

coordinates x = X/L, y = Y/B, the equations determining the deflection W = w(x,y) exp (iΩ T) are

$$\frac{\partial^2 w}{\partial y^2} + \epsilon^2 \frac{\partial^2 w}{\partial x^2} + \omega^2 w = 0$$

$$w(0,y) = W_0 \cos\{\pi y / 2G(0)\} \tag{2-1}$$

$$w(1,y) = w(x, \pm G(x)) = 0$$

The slenderness parameter ϵ = B/L is small. The dimensionless frequency $\omega = \frac{\Omega B}{c}$, where Ω is the excitation frequency, and c is the phase velocity in the membrane. The shape function G(x) is such that $G'(x) > 0$. A two variable expansion in the longitudinal coordinate is assumed in the form

$$w(x,y) = \gamma(\epsilon)\{w_0(x,\tilde{x},y) + \epsilon w_1(x,\tilde{x},y) + \cdots\} \tag{2-2}$$

where x is the 'slow' variable, and $\tilde{x}$ the 'fast' variable defined by

$$\frac{d\tilde{x}}{dx} = \frac{1}{\epsilon} k(x) \tag{2-3}$$

The function k(x) plays a central role in the theory and must be determined as part of the solution. The longitudinal derivatives are computed by treating x and $\tilde{x}$ as formally independent, so for example

$$\frac{\partial w}{\partial x} = \{\frac{\partial w_0}{\partial x} + \frac{1}{\epsilon} k(x) \frac{\partial w_0}{\partial \tilde{x}} + k(x) \frac{\partial w_1}{\partial \tilde{x}} + \epsilon \frac{\partial w_1}{\partial x} + \cdots\}\gamma(\epsilon)$$

With this expansion procedure, the dominant differential equations are

$$L w_0 \equiv \frac{\partial^2 w_0}{\partial y^2} + k^2(x) \frac{\partial^2 w_0}{\partial \tilde{x}^2} + \omega^2 w_0 = 0 \tag{2-4}$$

$$L w_1 = -k'(x) \frac{\partial w_0}{\partial \tilde{x}} - 2k(x) \frac{\partial^2 w_0}{\partial \tilde{x} \partial x} \tag{2-5}$$

Consider oscillatory solutions in $\tilde{x}$ in the separable form

$$w_0(x,\tilde{x},y) = \{a(x)\sin\tilde{x} + b(x)\cos\tilde{x}\}\eta(y;x) \tag{2-6}$$

An equation for $\eta(y;x)$ then follows from Eq. (2-4)

$$\frac{\partial^2 \eta}{\partial y^2} + \{\omega^2 - k^2(x)\}\eta = 0 \tag{2-7}$$

This equation is the analogue of a more complicated cross plane eigenvalue problem that occurs in the hydro-elastic problem. Here the cross plane problem is simply that of a string fixed at its endpoints $y = \pm G(x)$, and vibrating on an elastic foundation having the stiffness $k^2(x)$. For fixed ω, $k^2(x)$ is the eigenvalue. The appropriate solution of (2-7) consistent with all the boundary conditions is

$$\eta(y;x) = \cos\{\omega^2 - k^2(x)\}^{\frac{1}{2}} y \tag{2-8}$$

provided

$$\omega^2 - k^2(x) = \{\pi/2G(x)\}^2 \tag{2-9}$$

Since $k(x)$ can be interpreted as a local wavenumber, for fixed x Eq. (2-9) is just the dispersion relation for the particular cross plane mode being considered. More importantly, for the present problem, Eq. (2-9) tells how, for fixed ω, the local wavenumber changes with position. For monotonic increasing $G(x)$ the wavelength decreases monotonically. These waves can exist only if $\omega > \pi/2G(x)$. The slowly varying amplitude functions $a(x)$ and $b(x)$ in Eq. (2-6) are most simply determined from a solvability condition applied to the equation for w_1. The integral

$$\int_{-G(x)}^{G(x)} \eta(y;x)\, L w_1\, dy$$

can be put in the form

$$\int_{-G(x)}^{G(x)} \eta\{\omega^2 - k^2(x) - [\pi/2G(x)]^2\} w_1\, dy$$

and hence vanishes by Eq. (2-9). To get this form we require $\frac{\partial^2 w_1}{\partial \tilde{x}^2} = -w_1$ (i.e., w_1 has the same longitudinal wavelength structure as w_0), integrate by parts, and apply the boundary conditions $w_1(x, \tilde{x}, \pm G(x)) = 0$. The same integral over the right hand side of Eq. (2-5) then also vanishes. The coefficient of $\cos \tilde{x}$ after integration must vanish which yields an equation for $a(x)$

$$2k(x)G(x)a'(x)+[k(x)G(x)]'a(x)=0 \tag{2-10}$$

while the same equation results for b(x) from the coefficient of $\sin\tilde{x}$. Thus

$$a(x)=\frac{A}{\sqrt{k(x)G(x)}} \quad ; \quad b(x)=\frac{B}{\sqrt{k(x)G(x)}} \tag{2-11}$$

with A and B being arbitrary constants of integration. Collecting the results so far we have

$$w_0=\frac{1}{\sqrt{k(x)G(x)}}\cos\frac{\pi y}{2G(x)}\left\{A\sin\tilde{x}+B\cos\tilde{x}\right\} \tag{2-12}$$

with

$$\tilde{x}(x)=\epsilon^{-1}\int_c^x k(\xi)d\xi$$

This expression is uniformly valid in an interval where $k(x)>0$. For high enough frequency, $\omega>\frac{\pi}{2G(0)}$, Eq. (2-12) can be used to find the response of the entire length of the membrane. For this case the boundary conditions on the ends of the membrane at x = 0,1 determine the constants A and B, giving the result

$$w_0=\sqrt{\frac{k(0)G(0)}{k(x)G(x)}}\;\frac{\cos[\pi y/2G(x)]}{\sin\tilde{x}(1)}\sin[\tilde{x}(x)-\tilde{x}(1)] \tag{2-13}$$

$$\tilde{x}(x)=\epsilon^{-1}\int_0^x k(\xi)d\xi$$

Resonances occur at those frequencies satisfying $\tilde{x}(1)=n\pi$, or

$$\int_0^1 \{\omega^2 - [\pi/2G(x)]^2\}^{\frac{1}{2}} dx = \epsilon n\pi \qquad (2\text{-}14)$$

where the n are consecutive integers running from the nearest integer which is larger than

$$\epsilon^{-1} \int_0^1 \{(\tfrac{1}{G(0)})^2 - (\tfrac{1}{G(x)})^2\}^{\frac{1}{2}} dx$$

Exponential solutions in x also exist for Eq. (2-4). If instead of Eq. (2-6) we take

$$w_0(x,\tilde{x},y) = \{c(x)e^{\tilde{x}} + d(x)e^{-\tilde{x}}\}\,\eta(y;x) \qquad (2\text{-}15)$$

then both (2-8) and (2-9) follow if $g^2(x)$ replaces $-k^2(x)$. Differential equations for the amplitude functions c(x) and d(x) follow from the same solvability condition, the only difference being that we require $\frac{\partial^2 w_1}{\partial \tilde{x}^2} = w_1$. Then both c(x) and d(x) satisfy Eq. (2-10) with g(x) replacing k(x). The result is

$$c(x) = \frac{C}{\sqrt{g(x)G(x)}} \; ; \; d(x) = \frac{D}{\sqrt{g(x)G(x)}} \qquad (2\text{-}16)$$

where again C and D are arbitrary constants of integration. The result for the deflection is

$$w_0 = \frac{1}{\sqrt{g(x)G(x)}} \cos\frac{\pi y}{2G(x)} \{Ce^{\tilde{x}} + De^{-\tilde{x}}\} \qquad (2\text{-}17)$$

where

$$\tilde{x}(x) = \epsilon^{-1}\int_c^x g(\xi)d\xi \; ; \; g^2(x) = \{\pi/2G(x)\}^2 - \omega^2$$

For low enough frequency, $\omega < \pi/2G(1)$, Eq. (2-17) can be used to find a uniformly valid response over the entire length of the membrane. Satisfying the end conditions gives

$$w_0 = -W_0 \sqrt{\frac{g(0)G(0)}{g(x)\,G(x)}}\ \frac{\cos[\pi y/2G(x)]}{\sinh \tilde{x}(1)} \sinh[\tilde{x}(x)-\tilde{x}(1)] \qquad (2\text{-}18)$$

$$\tilde{x}(x) = \epsilon^{-1}\int_0^x g(\xi)\,d\xi$$

This response is sub-resonant and decays rapidly from the excitation end.

The frequency range $\pi/2G(0) < \omega < \pi/2G(1)$ is more interesting in that both forms, Eq. (2-12) and Eq. (2-17) must be used, and a connection must be made through the transition point where $\omega = \pi/2G(x_t)$. The amplitude of both these forms tends to infinity at the transition point. Near x_t we use an expansion for the transition layer

$$w(x,y) = \delta(\epsilon)\{w_0^*(x^*,y) + \epsilon^{\frac{2}{3}} w_1^*(x^*,y) + \cdots\}$$
$$x^* = \epsilon^{-2/3}(x - x_t) \qquad (2\text{-}19)$$

and consider the limit $\epsilon \to 0$ with x^* fixed. The dominant equations are

$$L^* w_0^* \equiv \frac{\partial^2 w_0^*}{\partial y^2} + \omega^2 w_0^* = 0 \qquad (2\text{-}20)$$

$$L^* w_1^* = -\frac{\partial^2 w_0^*}{\partial x^{*2}} \qquad (2\text{-}21)$$

The zero deflection boundary condition on the curved sides must also be expanded near x_t

$$w_0^*(x^*, G(x_t + \epsilon^{2/3}x^*)) + \epsilon^{2/3} w_1^*(x^*, G(x_t + \epsilon^{2/3}x^*)) + \cdots = 0$$

or

$$w_0^*(x^*, G(x_t)) + \epsilon^{2/3}\{x^* G'(x_t)\, w_{0_y}^*(x^*, G(x_t)) + w_1^*(x^*, G(x_t))\} + \cdots = 0$$

which gives

$$w_0^*(x^*, G(x_t)) = 0 \tag{2-22}$$

$$w_1^*(x^*, G(x_t)) = -x^* G'(x_t)\, w_{0_y}^*(x^*, G(x_t)) \tag{2-23}$$

Since $\omega = \pi/2G(x_t)$, the solution of (2-20) satisfying (2-22) is

$$w_0^*(x^*, y) = A^*(x^*)\cos[\pi y/2G(x_t)] = A^*(x^*)\,\eta(y; x_t) \tag{2-24}$$

Multiplying Eq. (2-21) by $\eta(y,x_t)$ and integrating over the width at fixed x gives an equation for $A^*(x^*)$

$$\frac{d^2 A^*(x^*)}{dx^{*2}} + \beta^3 x^* A^*(x^*) = 0 \tag{2-25}$$

$$\beta^3 = \pi^2 G'(x_t)/2G^3(x_t)$$

The general solution of Eq. (2-25) is

$$A^*(x^*) = E\,A_i(-\beta x^*) + F\,B_i(-\beta x^*) \tag{2-26}$$

where A_i and B_i are the standard Airy functions, and E,F constants which must be determined by matching to the oscillatory and exponential two variable expansions. Goodier and Fuller obtained Eq. (2-26) and determined the constants from the end conditions. However for a long narrow membrane the asymptotic matching procedure to be used here is preferable and leads to numerically different results. The matching is most conveniently carried out in terms of an intermediate variable

$$x_\sigma = \frac{x - x_t}{\sigma(\epsilon)} \qquad \epsilon^{2/3} \ll \sigma(\epsilon) \ll 1$$

The oscillatory, exponential, and transition layer expansions are written in terms of x_σ and the limits are taken as $\sigma(\epsilon)\to 0$ with x_σ fixed. Note that in terms of x_σ, $x^* = \sigma(\epsilon)\epsilon^{-2/3}x_\sigma \to -\infty$, and $x = x_t + \sigma x_\sigma \to x_t$. In the definition of x used in Eq. (2-17) we choose the lower limit of integration c = 0, so that the boundary condition at x = 0 is satisfied if

$$C + D = W_0\sqrt{g(0)G(0)} \tag{2-27}$$

Since

$$g(x) = g(x_t + \sigma x_\sigma) \simeq \sqrt{-\beta^3\sigma x_\sigma} + \cdots \qquad (x_\sigma < 0)$$

$$\tilde{x}(x) = \tilde{x}(x_t + \sigma x_\sigma) = \epsilon^{-1}\int_0^{x_t+\sigma x_\sigma} g(\xi)\,d\xi \simeq \tilde{x}_1(x_t) + \frac{2}{3\epsilon}(-\beta\sigma x_\sigma)^{3/2} + \cdots$$

the intermediate limit of Eq. (2-17) is

$$w(x_\sigma, y) \simeq \frac{\cos[\pi y/2G(x_t)]}{G(x_t)^{1/2}(-\beta^3\sigma x_\sigma)^{1/4}}\Big\{C\exp\Big[\tilde{x}_1(x_t) + \frac{2}{3\epsilon}(-\beta\sigma x_\sigma)^{3/2}\Big] + D\exp\Big[-\tilde{x}_1(x_t) - \frac{2}{3\epsilon}(-\beta\sigma x_\sigma)^{3/2}\Big]\Big\} + \cdots \tag{2-28}$$

Using the asymptotic expansions of the Airy functions for large positive arguments we obtain the intermediate limit of the transition layer expansion

$$w(x_\sigma, y) \simeq \frac{\epsilon^{\frac{1}{6}}\delta(\epsilon)\cos[\pi y/2G(x_t)]}{\sqrt{\pi}\,(-\beta\sigma x_\sigma)^{1/4}}\Big\{\frac{1}{2}E\exp\Big[-\frac{2}{3\epsilon}(-\beta\sigma x_\sigma)^{3/2}\Big] + F\exp\Big[\frac{2}{3\epsilon}(-\beta\sigma x_\sigma)^{3/2}\Big]\Big\} + \cdots \tag{2-29}$$

The two expansions, Eq. (2-28) and Eq. (2-29) asymptotically match if $\delta(\epsilon) = \epsilon^{-\frac{1}{6}}$ and

$$E = 2\left(\frac{2\pi}{G'(x_t)}\right)^{\frac{1}{6}} D e^{-\tilde{x}_1(x_t)} \tag{2-30}$$

$$F = \left(\frac{2\pi}{G'(x_t)}\right)^{\frac{1}{6}} C e^{\tilde{x}_1(x_t)} \tag{2-31}$$

For the oscillatory expansion we choose x to be defined as in (2-12) with c = 1, then B = 0 will satisfy zero deflection at x = 1. Since

$$k(x) = k(x_t + \sigma x_\sigma) \simeq \sqrt{\beta^3 \sigma x_\sigma} + \cdots \qquad (x_\sigma > 0)$$

$$\tilde{x}(x) = \tilde{x}(x_t + \sigma x_\sigma) = \epsilon^{-1} \int_1^{x_t + \sigma x_\sigma} k(\xi)\, d\xi \simeq \tilde{x}_3(x_t) + \frac{2}{3\epsilon} (\beta \sigma x_\sigma)^{3/2} + \cdots$$

the intermediate limit of Eq. (2-12) is

$$w(x_\sigma, y) \simeq \frac{\gamma(\epsilon) A \cos[\pi y / 2G(x_t)]}{\sqrt{G(x_t)}\ (\beta^3 \sigma x_\sigma)^{\frac{1}{4}}} \Big\{ \sin \tilde{x}_3(x_t) \cos \frac{2}{3\epsilon} (\beta \sigma x_\sigma)^{3/2}$$

$$+ \cos \tilde{x}_3(x_t) \sin \frac{2}{3\epsilon} (\beta \sigma x_\sigma)^{3/2} \Big\} + \cdots \qquad (2\text{-}32)$$

Now use the asymptotic expansions of the Airy functions for large negative argument to obtain the intermediate limit of the transition layer expansion

$$w(x_\sigma, y) \simeq \frac{\cos[\pi y / 2G(x_t)]}{\sqrt{\pi}\ (\beta \sigma x_\sigma)^{\frac{1}{4}}} \Big\{ E \sin\big[\tfrac{2}{3\epsilon} (\beta \sigma x_\sigma)^{3/2} + \pi/4\big]$$

$$+ F \cos\big[\tfrac{2}{3\epsilon} (\beta \sigma x_\sigma)^{3/2} + \pi/4\big] \Big\} + \cdots \qquad (2\text{-}33)$$

Eq. (2-32) is identical with Eq. (2-33) if $\gamma(\epsilon) = 1$ and

$$E - F = \sqrt{2} \left(\frac{2\pi}{G'(x_t)} \right)^{\frac{1}{6}} A \cos \tilde{x}_3(x_t) \qquad (2\text{-}34)$$

$$E + F = \sqrt{2} \left(\frac{2\pi}{G'(x_t)} \right)^{\frac{1}{6}} A \sin \tilde{x}_3(x_t) \qquad (2\text{-}35)$$

The unknown constants can now be determined from the equations (2-27), (2-30), (2-31), (2-34) and (2-35). Solving the system gives

$$A\Delta = -2\sqrt{2} \sec \tilde{x}_3(x_t)$$

$$C\Delta = 2 \exp[-\tilde{x}_1(x_t)] (1 - \tan \tilde{x}_3(x_t))$$

$$D\Delta = -\exp[\tilde{x}_1(x_t)] (1 + \tan \tilde{x}_3(x_t))$$

$$E\Delta = 2 [2\pi / G'(x_t)]^{1/6} (1 + \tan \tilde{x}_3(x_t)) \qquad (2\text{-}36)$$

$$F\Delta = 2[2\pi/G'(x_t)]^{1/6}(1-\tan \tilde{x}_3(x_t))$$

$$\Delta = \frac{2\exp[-\tilde{x}_1(x_t)](1-\tan \tilde{x}_3(x_t)) - \exp[\tilde{x}_1(x_t)](1+\tan \tilde{x}_3(x_t))}{W_0[g(0)G(0)]^{1/2}}$$

These results indicate that the deflection is extremely small, $O(e^{-\frac{1}{\epsilon}})$, away from the excitation end, provided the excitation frequency is such that $-\tan \tilde{x}_3(x_t)$ is not very close to unity. On the other hand if it happens that tan x_3 $(x_t) = -1$ exactly, then the deflection is exponentially large, $O(e^{\frac{1}{\epsilon}})$. The resonant frequencies are determined from the condition $\Delta = 0$.

It is of some interest to compare numerical results obtained from this theory with the theory of Goodier and Fuller. We consider the specific planform geometry applicable to the cochlea: $G(x) = (1+5x)/6$ and $\epsilon^{-1} = 120$. In their theory the transition point is arbitrarily positioned at $x_t = 1/2$. That is, they consider only those combinations of excitation frequency and linear planform taper so that $x_t = 1/2$. In the present example, $x_t = 1/2$ if $\omega = \pi/2G(x_t) = 2.69$. Their theory gives for $W_0 = 1$ the center line deflection at the transition point 1.10×10^{-56}. Our theory gives the corresponding deflection -0.83×10^{-94}. In this example both deflections are $O(e^{-\frac{1}{\epsilon}})$ (and not experimentally detectable), since $\omega = 2.69$ is not close to a resonant frequency. But nevertheless the ratio of the two results is 38 orders of magnitude and is indicative of the sensitivity of the calculation for small ϵ.

This problem illustrates the use of the two variable expansion procedures, transition or turning points, and matching to a transition layer expansion which is needed to connect the two variable solutions. For a tapered membrane or plate immersed in a bounded fluid, the transition point phenomenon can also occur, but only for transverse mode shapes having zero mean deflection.

III. A Two Dimensional Problem

Consider now the problem illustrated in Figure 2 where a partition having distributed mass, stiffness, and damping is immersed in a two dimensional inviscid fluid which is bounded by a rigid wall.

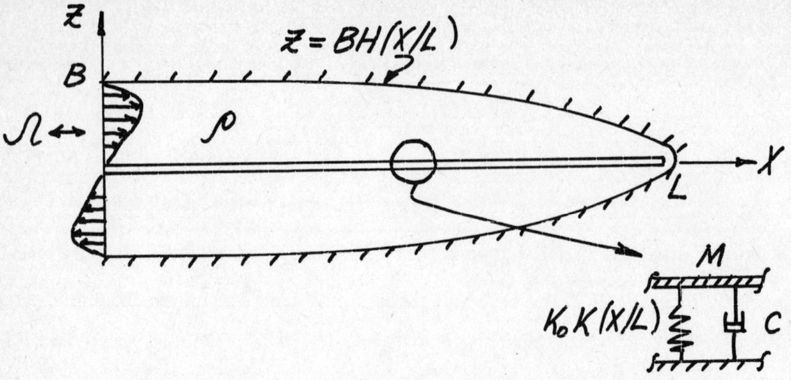

Figure 2. Geometry and Parameters of Two Dimensional Problem

An inviscid, incompressible fluid of density ρ is divided by a partition along the X axis into two symmetrical regions. The upper rigid boundary is located at Z = BH(X/L) where H(X/L) is a slowly varying dimensionless shape function. The stiffness per unit area of partition is $K_0K(X/L)$ (dyne/cm^3) where K(X/L) is a dimensionless stiffness function which is a decreasing function of its argument. M (gm/cm^2) is the mass of the partition per unit area, and C (dyne-sec/cm^3) is the internal damping constant of the partition per unit area.

The system is forced at X = 0, $0 \leq Z \leq B$ by a flux of fluid which oscillates with frequency Ω. This problem has been considered by quite a few authors (at least for H'(x) = 0) who have computed the solution using a variety of methods. The reader is referred to the papers of Allen (1977), Allen and Sondhi (1979), deBoer (1979), Lesser and Berkley (1972), Neely (1980), Ranke (1950), Siebert (1974), Sondhi (1978) and van Dijk (1976). The emphasis here is on the WKB approach which was also used by Steele and Taber (1979), Viergever (1980), and Lighthill (1980). Here we obtain some further analytical results which are applicable to the case of small damping. In the cochlea the relative importance of fluid damping to partition damping is unknown, but some agreement with experiment has been obtained using only a small partition damping.

The dependent variables are the deflection of the partition W(X,T), and the velocity potential $\Phi(X,Z,T)$. The velocity field is then given by $\vec{U} = \nabla\Phi$, and the pressure by $P = -\rho\partial\Phi/\partial T$ for small motions. The independent variables are scaled as

$$x = \frac{X}{L} \; ; \; z = \frac{Z}{\lambda_0(\Omega)} \; ; \; t = \Omega T$$

where $\lambda_0 = K_0/\rho\Omega^2$ is a reference wavelength. Then writing

$$W(X,T) = W_A\, w(x) e^{it}$$
$$\Phi(X,Z,T) = \Phi_A\, \varphi(x,z) e^{it}$$

with $\Phi_A = \lambda_0 \Omega W_A$ we obtain the dimensionless equations and boundary conditions

$$\frac{\partial^2\varphi}{\partial z^2} + \epsilon^2 \frac{\partial^2\varphi}{\partial x^2} = 0 \tag{3-1}$$

$$\{K(x) - m + i\beta\} w(x) = 2i\,\varphi(x,0^+) \tag{3-2}$$

$$\frac{\partial\varphi}{\partial z} = iw \quad \text{on } z = 0 \tag{3-3}$$

$$\frac{\partial\varphi}{\partial z} = \frac{\epsilon^2}{\mu} H'(x) \frac{\partial\varphi}{\partial x} \quad \text{on } z = \frac{1}{\mu} H(x) \tag{3-4}$$

$$\frac{\partial\varphi}{\partial x} = \frac{1}{\epsilon} u(z) \quad \text{at} \quad x = 0 \tag{3-5}$$

$$\varphi(1,z) = 0 \tag{3-6}$$

The equations express, in turn, incompressibility of the fluid, balance between force and inertia of the partition, the vertical component of fluid velocity on the partition equals the partition velocity, there is no flux of fluid through the rigid wall, the horizontal velocity profile u(z) is specified at the stapes (x=0), and pressure is equalized in the two chambers at the helicotrema (x = 1). Four parameters appear in the problem

$$\epsilon = \frac{\lambda_0}{L} \; ; \; \mu = \frac{\lambda_0}{B} \; ; \; m = \frac{M}{\rho\lambda_0} \; ; \; \beta = \frac{C\Omega}{K_0}$$

and we seek a solution initially with $\epsilon \to 0$ (Ω large) and μ, m, β fixed. Consider now the WKB expansions

$$\left\{ \begin{array}{l} \varphi(x,z) \\ w(x) \end{array} \right\} = e^{-i\tilde{x}} \left\{ \begin{array}{l} \varphi_0(x,z) + \epsilon \varphi_1 + \cdots \\ w_0(x) + \epsilon w_1 + \cdots \end{array} \right. \tag{3-7}$$

$$\tilde{x}(x) = \epsilon^{-1} \int_0^x k(x)\,dx$$

where k(x) is the phase function to be found. These expansions yield the following set of dominant equations

$$\frac{\partial^2 \varphi_0}{\partial z^2} - k^2(x)\varphi_0 = 0 \tag{3-8}$$

$$\{K(x) - m + i\beta\} w_0 = 2i\varphi_0(x, 0^+) \tag{3-9}$$

$$\frac{\partial \varphi_0}{\partial z} = i w_0 \quad \text{on } z = 0 \tag{3-10}$$

$$\frac{\partial \varphi_0}{\partial z} = 0 \quad \text{on } z = \frac{1}{\mu} H(x) \tag{3-11}$$

The function

$$\varphi_0(x,z) = -\frac{i w_0(x) \cosh\left[k(x)\left(\frac{H(x)}{\mu} - z\right)\right]}{k(x) \sinh\left[k(x) H(x)/\mu\right]}, \quad z \geq 0 \tag{3-12}$$

satisfies (3-8), (3-10) and (3-11). Substitution of (3-12) into (3-9) then yields the dispersion relation

$$\{K(x) - m + i\beta\} k(x) = 2 \coth\left[k(x) H(x)/\mu\right] \tag{3-13}$$

Solutions of this equation determine the phase function k(x), which will be further discussed later. To find $w_0(x)$ in (3-12) we consider the $O(\epsilon)$ equations

$$\frac{\partial^2 \varphi_1}{\partial z^2} - k^2 \varphi_1 = 2ik \frac{\partial \varphi_0}{\partial x} + ik' \varphi_0 \tag{3-14}$$

$$\{K(x) - m + i\beta\} w_1 = 2i \varphi_1(x, 0^+) \tag{3-15}$$

$$\frac{\partial \varphi_1}{\partial z} = i w_1 \quad \text{on } z = 0 \tag{3-16}$$

$$\frac{\partial \varphi_1}{\partial z} = -i \frac{k}{\mu} H'(x) \varphi_0 \quad \text{on } z = \frac{1}{\mu} H(x) \tag{3-17}$$

These equations together with the first order equations imply an important integral relation

$$k(x) \int_0^{H(x)/\mu} \varphi_0^2(x,z)\, dz = \text{constant} \tag{3-18}$$

This has the physical interpretation that the rate of working (the product of pressure and longitudinal velocity) is constant over every cross section. It is interesting that (3-18) holds even with dissipation in the partition. To prove (3-18) multiply (3-14) by φ_0 and integrate over a cross section

$$\int_0^{H(x)/\mu} \varphi_0 \left(\frac{\partial^2 \varphi_1}{\partial z^2} - k^2 \varphi_1 \right) dz = 2ik \int_0^{H/\mu} \varphi_0 \frac{\partial \varphi_0}{\partial x}\, dz + ik' \int_0^{H/\mu} \varphi_0^2\, dz$$

Integrating the left hand side by parts and using (3-8) leaves the form

$$\left(\varphi_0 \frac{\partial \varphi_1}{\partial z} - \varphi_1 \frac{\partial \varphi_0}{\partial z} \right) \Big|_0^{H/\mu}$$

Evaluation of the boundary values then leads to

$$2k \int_0^{H/\mu} \varphi_0 \frac{\partial \varphi_0}{\partial x}\, dz + k' \int_0^{H/\mu} \varphi_0^2\, dz + k\, \varphi_0^2(x, H/\mu) H'(x)/\mu = 0$$

which is the x-derivative of (3-18). The partition displacement $w_0(x)$ can now be found from (3-12) and (3-18)

$$w_0(x) = \frac{C_0 k \sinh(kH/\mu)}{[Hk/\mu + \frac{1}{2}\sinh(2kH/\mu)]^{1/2}} \tag{3-19}$$

where C_o is an arbitrary constant. The solution obtained so far is equivalent to that found by Steele and Taber (1979) using the time-averaged Lagrangian formalism.

The end conditions (3-5) and (3-6) must still be satisfied to complete the solution. Fortunately, as will be demonstrated, the dispersion relation (3-13) has (for fixed x) an infinite number of roots $k_n(x)$, n = 1, 2, . . . For each $k_n(x)$ there are corresponding modes $\varphi_{o_n}(x, z)$ and $w_{o_n}(x)$ given by (3-12) and (3-19). A linear combination of the φ_{o_n} can then be used to satisfy (3-15), the forcing condition. In the usual way an orthogonality condition can be demonstrated for any two distinct modes φ_{o_n} and φ_{o_m} ($n \neq m$). Since $\partial^2\varphi_{o_n}/\partial z^2 = k_n^2\varphi_{o_n}$ and $\partial^2\varphi_{o_m}/\partial z^2 = k_m^2\varphi_{o_m}$ then

$$(k_n^2 - k_m^2)\int_0^{H/\mu} \varphi_{o_n}\varphi_{o_m}\,dz = \int_0^{H/\mu} \{\varphi_{o_m}\frac{\partial^2\varphi_{o_n}}{\partial z^2} - \varphi_{o_n}\frac{\partial^2\varphi_{o_m}}{\partial z^2}\}\,dz$$

The right hand side can be shown to vanish after integration by parts and evaluation of boundary terms. Since $k_n \neq k_m$ for $n \neq m$ we obtain the orthogonality condition

$$\int_0^{H/\mu} \varphi_{o_n}\varphi_{o_m}\,dz = 0 \tag{3-20}$$

It has been customary in previous solutions to this problem to say that a boundary condition such as (3-6) will automatically be satisfied to a high degree of approximation using only right running waves, i.e. the reflected left running wave will have negligible amplitude. This may in fact be the case depending on the parameters, but since it is a numerical question it is preferable to first satisfy the boundary condition at x = 1 and then see from the solution if the right running waves are dominant. We note from (3-13) that if $k_n(x)$ is a root, then so is $-k_n(x)$. Thus the dispersion relation produces both right and left running waves. This fact can be used to satisfy (3-6) exactly by taking

$$\varphi(x,z) = \sum_n \alpha_n \varphi_{0n}(x,z) \left\{ \frac{e^{i[\tilde{x}_n - \tilde{x}_n(1)]} - e^{-i[\tilde{x}_n - \tilde{x}_n(1)]}}{2\cos \tilde{x}_n(1)} \right\} + O(\epsilon)$$

$$\tilde{x}_n(x) = \epsilon^{-1} \int_0^x k_n(x)dx \qquad (3\text{-}21)$$

It will be seen that all the solutions of the dispersion relation (3-13) lie in the second and fourth quadrants of the complex plane. The sum indicated by (3-21) is understood to be over the fourth quadrant roots. Now the forcing function (3-5) can be satisfied to first order by using (3-21) and the orthogonality condition (3-20) to determine

$$\alpha_n = \int_0^{H/\mu} u(z)\, \varphi_{0n}(0,z)dz \qquad (3\text{-}22)$$

where the constant C_0 in (3-19) has been set to $\sqrt{2i}$ which normalizes the eigenfunctions in such a way that

$$k_n(x) \int_0^{H/\mu} \varphi_{0n}^{\,2}(x,z)dz = -i \qquad (3\text{-}23)$$

with

$$\varphi_{0n}(x,z) = \frac{\sqrt{-2i}\, \cosh[k_n(H/\mu - z)]}{\{k_n H/\mu + \frac{1}{2}\sinh 2k_n H/\mu\}^{\frac{1}{2}}} \qquad (3\text{-}24)$$

The partition displacement follows from (3-21) and (3-2)

$$w(x) = \sqrt{2i} \sum_n \alpha_n \frac{k_n \sinh k_n H/\mu}{\{k_n H/\mu + \frac{1}{2}\sinh 2k_n H/\mu\}^{\frac{1}{2}}} \left\{ \frac{e^{i[\tilde{x}_n - \tilde{x}_n(1)]} - e^{-i[\tilde{x}_n - \tilde{x}_n(1)]}}{2\cos \tilde{x}_n(1)} \right\}$$
$$+ O(\epsilon) \qquad (3\text{-}25)$$

which completes the formal first order solution.

Finding the roots $k_n(x)$ of (3-13) is of course the difficult aspect of the problem, so it is worthwhile to discuss some of their properties. First let $k(x) = \xi(x) + i\eta(x)$ with the subscript n understood. Then separation of (3-13) into real and imaginary parts leads to the coupled system

$$\{K(x) - m\}\xi - \beta\eta = \frac{2\sinh 2\frac{H}{\mu}\xi}{\cosh 2\frac{H}{\mu}\xi - \cos 2\frac{H}{\mu}\eta} \tag{3-26}$$

$$\beta\xi + \{K(x) - m\}\eta = -\frac{2\sin 2\frac{H}{\mu}\eta}{\cosh 2\frac{H}{\mu}\xi - \cos 2\frac{H}{\mu}\eta} \tag{3-27}$$

These equations determine the real and imaginary parts of the wavenumber as a function of distance along the partition. A relation determining the root loci, i.e. η vs. ξ is also useful. Let $\tilde{\xi} = \beta\xi/2$ and $\tilde{\eta} = \beta\eta/2$ and equate the factor $\{K(x) - m\}$ in each of the above equations to obtain

$$(\tilde{\eta}^2 + \tilde{\xi}^2)(\cosh\kappa\tilde{\xi} - \cos\kappa\tilde{\eta}) + \tilde{\eta}\sinh\kappa\tilde{\xi} + \tilde{\xi}\sin\kappa\tilde{\eta} = 0 \tag{3-28}$$

where $\kappa = 4H(x)/\mu\beta$. At high enough frequency and/or small enough damping κ turns out to be large, which will be the basis for further asymptotic approximations.

It is readily seen from (3-13) that for any x and nonzero β all roots are fully complex, i.e. $\xi = 0$ and/or $\eta = 0$ is not possible. Furthermore, no roots lie in the first quadrant of the complex plane. This can be seen from (3-28). For $\xi, \eta > 0$ the first term is positive so it is sufficient to show the magnitude of the second term exceeds the magnitude of the third term

$$\tilde{\eta}\sinh\kappa\tilde{\xi} > \tilde{\xi}|\sin\kappa\tilde{\eta}|$$

which is easily verified since

$$\frac{|\sin\kappa\tilde{\eta}|}{\kappa\tilde{\eta}} < 1 < \frac{\sinh\kappa\tilde{\xi}}{\kappa\tilde{\xi}}$$

If no roots lie in the first quadrant then also no roots lie in the third quadrant since (3-28) is invariant to simultaneous sign changes in $\tilde{\xi}$ and $\tilde{\eta}$. Thus all existing roots of (3-13) lie in the second and fourth quadrants. A fourth quadrant root represents a damped right running wave with complex wave number $k(x) = \xi(x) + i\eta(x)$ ($\xi > 0$, $\eta < 0$). For every such $k(x)$, $-k(x)$ exists ($\xi < 0$, $\eta > 0$) and represents a damped left running wave.

Numerical solutions of (3-28) for the fourth quadrant roots have been carried out on a programmable hand calculator and are shown in Figure 3. The dominant propagating root (i.e. the root having the smallest $|\tilde{\eta}|$) is shown for different values of κ. κ is a constant for the case of constant chamber height and constant damping coefficient. As κ increases a significant portion of the curve approaches a semi-circle. This can be seen from (3-28) by noting for $\tilde{\xi} = 0(1)$, $\tilde{\eta} = 0(1)$, and $\kappa \gg 1$ then

$$\tilde{\eta}^2 + \tilde{\xi}^2 + \tilde{\eta} + O(e^{-\kappa}) = 0$$

Neglecting the exponentially small terms, this is a semi-circle having radius 1/2 and centered at $\tilde{\xi} = 0$, $\tilde{\eta} = -1/2$. This approximation breaks down when $\tilde{\xi}$ is small along the $\tilde{\eta}$ axis where the curve is oscillatory. The semi-circle limit physically represents the "deep water" or no top approximation. The existence of an infinite number of roots can also be demonstrated analytically for $\kappa \gg 1$. Consider small $\tilde{\xi}$ such that $\kappa\tilde{\xi} \ll 1$. Then the hyperbolic functions in (3-28) can be expanded for small arguments

$$(\tilde{\eta}^2 + \dots)\{1 + \tfrac{1}{2}(\kappa\tilde{\xi})^2 + \dots - \cos\kappa\tilde{\eta}\} + \tilde{\eta}\{\kappa\tilde{\xi} + \dots\} + \dots = 0$$

If $|\tilde{\eta}| > 1$, say, then a balance can be achieved only if $\cos\kappa\tilde{\eta} \simeq 1$, or $\kappa\tilde{\eta} \simeq -2n\pi$, where n is a large positive integer. Let $\kappa\tilde{\eta} = -2\pi n + \kappa\eta'$, where $\kappa\eta' \ll 1$. Then the dominant balance gives the circular orbits

$$\left(\tilde{\eta} + \frac{2\pi n}{\kappa}\right)^2 + \left(\tilde{\xi} - \frac{1}{2\pi n}\right)^2 = \left(\frac{1}{2\pi n}\right)^2$$

The approach to these curves is shown in Figure 3 for $\kappa = 40$. These roots correspond to the higher order modes needed to satisfy the flux condition at $x = 0$. Because these roots are characterized by large η, they represent waves which are rapidly damped away from $x = 0$. The relative importance of these roots depends on the shapes of the velocity profile at $x = 0$.

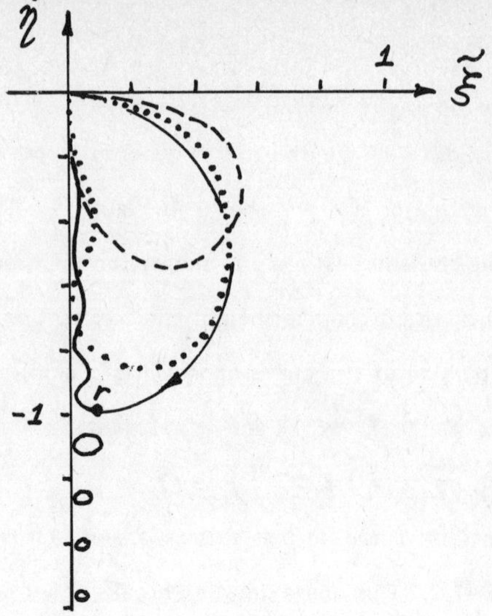

Figure 3. Root Loci (—— χ = 40; - - - χ = 5; ...... χ = 10)

The root loci plot of Figure 3 helps to clarify some aspects of the complex roots of the dispersion relation, but it provides no information concerning the x-dependence of the roots. This is what is actually needed to calculate the response of the system. This information must come from (3-26) and (3-27). Here we shall concentrate on the behavior of the solution near the region of maximal response, i.e. near the longitudinal station x_r, where $K(x_r) = \mathcal{M}$ the "resonance region." We define a new coordinate $x^* = (x - x_r)/\beta$ where for $\beta \ll 1$, $x^* = 0(1)$ in close proximity to the resonance location. Then $K(x) - \mathcal{M} = K(x_r + \beta x^*) - \mathcal{M} = \beta x^* K'(x_r) + 0(\beta^2)$. Then neglecting the $0(\beta^2)$ term which is the same approximation made by deBoer (1979), we obtain from (3-26) and (3-27) the simpler system

$$\tau U + V = -\chi \frac{\sinh U}{\cosh U - \cos V} \tag{3-29}$$

$$U - \tau V = -\chi \frac{\sin V}{\cosh U - \cos V} \tag{3-30}$$

where

$$\tau = -K'(x_r)x^* \; ; \; U = \kappa\tilde{\xi} \; ; \; V = \kappa\tilde{\eta} \tag{3-31}$$

Notice that since $K'(x_r) < 0$, $\tau > 0$ when $x > x_r$. Some simple asymptotic approximations can be made for $\kappa >> 1$. Consider the expansions

$$\begin{aligned} U &= \kappa U_0 + U_1 + \cdots \\ V &= \kappa V_0 + V_1 + \cdots \end{aligned} \tag{3-32}$$

which give the dominant equations for τ, U_0, V_0 all 0(1)

$$\begin{aligned} \tau U_0 + V_0 &= -1 \\ U_0 - \tau V_0 &= 0 \end{aligned} \tag{3-33}$$

so that

$$U_0 = \frac{-\tau}{1+\tau^2} \; ; \; V_0 = -\frac{1}{1+\tau^2} \tag{3-34}$$

This approximation gives explicit x-dependence of the real and imaginary parts of the complex wavenumber. It corresponds to the semi-circle or deep water approximation of the dominant root shown in Figure 3. It is valid only to the left of the resonance point $\tau < 0$, since the wavenumber must lie in the fourth quadrant of the complex plane. Nevertheless, it is a very useful approximation since it determines the dominant response. To continue past the resonance location we consider different expansions

$$\begin{aligned} U &= U_0^* + \frac{1}{\kappa}U_1^* + \cdots \\ V &= -\kappa + V_1^* + \cdots \end{aligned} \tag{3-35}$$

which give the dominant equations

$$\begin{aligned} 1 &= \frac{\sinh U_0^*}{\cosh U_0^* - \cos(V_1^*\kappa)} \\ \tau &= -\frac{\sin(V_1^*\kappa)}{\cosh U_0^* - \cos(V_1^*\kappa)} \end{aligned} \tag{3-36}$$

These equations can be solved explicitly for $U_o^*(\tau)$ and $V_1^*(\tau)$

$$U_o^* = (1/2)\log(1 + 4/\tau^2)$$

$$V_1^* = \varkappa + \tan^{-1}(2/\tau) \qquad (3\text{-}37)$$

The branch of the inverse tangent must be such that $V_1^* = 0(1)$. With this understanding (3-37) implies

$$\tilde{\xi} \sim \frac{1}{2\kappa}\log(1 + 4/\tau^2) + \cdots$$

$$\tilde{\eta} \sim -\frac{1}{\kappa}\{n\pi + \tan^{-1}(2/\tau)\} + \cdots \qquad (3\text{-}38)$$

where n is an integer and the principal branch of the inverse tangent is understood. In (3-38) $\tilde{\xi}$ has a logorithmic singularity at the resonance point τ=0, so (3-38) is valid only for $\tau > 0$. A further expansion is needed for small τ which will match to (3-34) on the left and (3-38) on the right. Further work is needed in this region, particularly to firmly establish which value of n to use in (3-38). This point is apparently related to numerical problems encountered in previous calculations. The calculations of Steele and Taber (1979) show a non-uniqueness in the region past resonance, which doesn't seem to converge to a definite answer as the step size was reduced. Viergever's (1980) results allow for a jump in the wavenumber across the resonance region. This would seem to locally invalidate the solution since the wavenumber must possess a derivative in the WKB formalism. Numerical experiments, at least, suggest that for the dominant root and $\varkappa >> 1$, n should be chosen in such a way that $|n\pi/\kappa|$ is less than but closest to unity. This procedure gives a continuous wavenumber, when the asymptotic formulas (3-34) are used for $\tau < 0$, (3-38) used for $\tau > 0$, and 'exact' numerical computations are used in the vicinity of $\tau = 0$. The 'exact' computations were done by uncoupling U and V in (3-29) and (3-30) yielding a single transcendental equation for either U or V, and then subsequent trial and error solution on a programmable hand calculator. The composite results are shown in Figure 4 for the dominant root at $\varkappa = 40$. Figure 4 helps in the interpretation of Figure 3. Far to the left of the resonance location the wavenumber is near the origin and then moves along the semicircle in the direction of the arrow to the point r which corresponds to the resonance location, or $\tau = 0$. For $\tau > 0$ the wavenumber stays close to the point r.

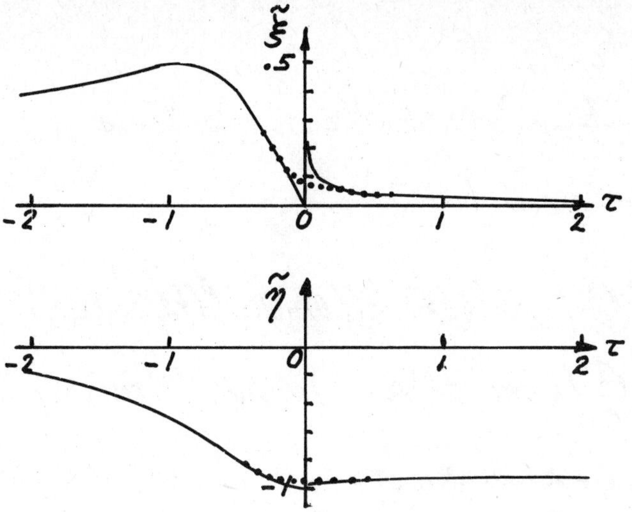

Figure 4. Wavenumber dependence on longitudinal station. Dominant root for χ = 40, n = 12.

Some convenient closed form analytical expressions for the partition deflection can be obtained based on (3-25) and the asymptotic approximations for the longitudinal dependence of the wavenumber (3-34) and (3-38). The resulting expressions can be written in the form

$$w(\tau) \sim \text{constant} \cdot a(\tau)\, e^{i\Theta(\tau)} \tag{3-39}$$

where $a(\tau)$, $\Theta(\tau)$ are the amplitude and phase functions near resonance. We find for $\tau < 0$

$$a(\tau) = \frac{e^{\frac{2}{\epsilon K'(\chi_r)} \tan^{-1}\tau}}{\sqrt{1+\tau^2}}$$

$$\Theta(\tau) = \tan^{-1}(1/\tau) - \frac{1}{\epsilon K'(\chi_r)} \log(1+\tau^2) \tag{3-40}$$

and for $\tau > 0$

$$a(\tau) = \frac{(a^2+b^2)^{\frac{1}{2}}}{(c^2+d^2)^{\frac{1}{4}}} e^{\frac{2}{\epsilon\kappa K'(x_r)}(\tau\theta + 2l)}$$

$$\Theta(\tau) = \frac{2}{\epsilon\kappa K'(x_r)}(\tau l + 2\tan^{-1}(\tau/2)) + \tan^{-1}(b/a) - \frac{1}{2}\tan^{-1}(d/c) \qquad (3\text{-}41)$$

where

$$a = \frac{\sqrt{2}}{\kappa}\{l \sinh(\tfrac{1}{2}l)\cos(\tfrac{1}{2}\theta) - \theta\cosh(\tfrac{1}{2}l)\sin(\tfrac{1}{2}\theta)\}$$

$$b = \frac{\sqrt{2}}{\kappa}\{l\cosh(\tfrac{1}{2}l)\sin(\tfrac{1}{2}\theta) + \theta\sinh(\tfrac{1}{2}l)\cos(\tfrac{1}{2}\theta)\}$$

$$c = l + 2(\cos n\pi)/(\tau^2+4)$$

$$d = -\{\theta + 2(\cos n\pi)(\tau^2+2)/\tau(\tau^2+4)\}$$

with

$$l = \frac{1}{2}\log(1+4/\tau^2)$$

$$\theta = n\pi + \tan^{-1}(2/\tau)$$

These expressions assume that only the dominant propagating root has been excited and that the reflected left running waves are negligible. These expressions lose some accuracy in the range $|\tau| \lesssim 0.1$. Also, these expressions should not be used for very large τ, since the stiffness function has been approximated by the first term of its Taylor series development.

A "snapshot" of some partition waveforms is shown in Figure 5. On a linear plot the wave is effectively cut off past resonance ($\tau > 0$), so only (3-40) is needed. The waveforms depend only on one parameter $1/\epsilon K'(x_r)$. For the special case of $K(x) = \exp(-\alpha Lx)$, which is usually considered by other authors, $1/\epsilon K'(x_r) = -\rho/M\alpha$ independent of frequency. The waves are shown for three different values of this parameter. The case $\rho M/\alpha = 10/3$ was used by Steele and Taber (1979) who compared their calculations with those of Neely (1980). As can be seen, the waveforms are quite sensitive to this parameter. The abcissa can be converted to physical lengths using $X - X_r = (C/\alpha M\Omega)\tau$, which is frequency dependent. Shown are 1 mm lengths at different frequencies using the parameters C = 200 dyne-sec/cm^3, M = .15 gm/cm^2, α = 2/cm.

In Figure 6 the amplitude function has been plotted on a logarithmic decibel (db) scale defined here as 20 $Log_{10}(|a(\tau)/a_{max}|)$. The phase is plotted on a linear scale, where we have taken the location of the maximum amplitude to correspond to zero phase. Computations are carried out for an excitation frequency of 4.5 KHZ with the above parameters. These calculations seem to agree with the improved calculations of Steele (1981).

Physiological amplitude and phase data have been attempted to be explained by models of this type, which are referred to as point-impedance models. Viergever (1981) reviews some of these efforts. Our feeling is that the underlying mechanics can be improved upon, which brings us to the next problem.

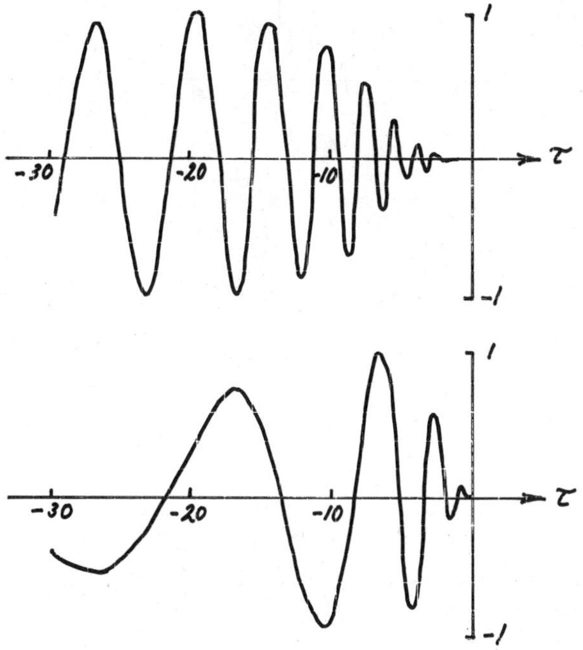

Figure 5. Partition Waveforms

The parameter $1/\varepsilon K'(x_r)$ takes on the values -10, -10/3, in going from top to bottom. Eqs. (3-39) and (3-40) are used to compute these curves.

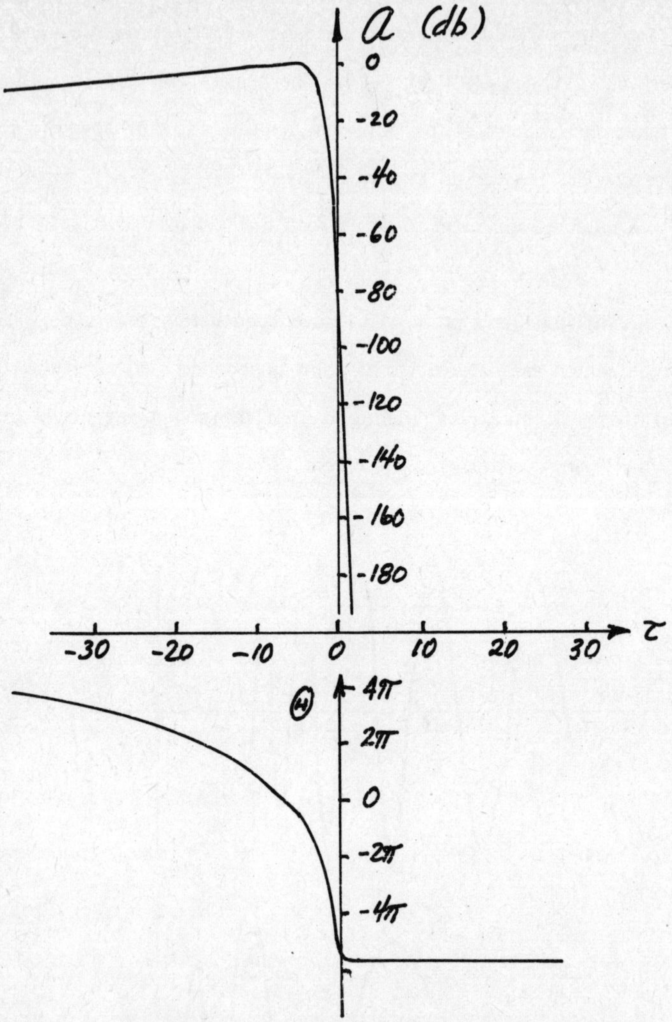

Figure 6. Amplitude and Phase of Partition

Eqs. (3-41) are used to compute this curve at 4.5 KHZ, $1/\varepsilon K'(x_r) = -10/3$, $\kappa = 56.55$, $n = 18$.

IV. A Three Dimensional Problem

We now turn to another mathematical representation of cochlear mechanics which is more difficult than the problem of Section III in that here we attempt to deal with the essential geometrical constraints of the cochlea. In addition, the fluid and the basilar membrane are given more realistic physical properties.

The hydroelastic system shown in Figure 7 consists of a slender (uncoiled) rigid tube having variable cross sectional area and filled with a viscous, incompressible, Newtonian fluid. The tube is divided lengthwise into two chambers by an interior surface, part of which is rigid (spiral ligament) and part elastic (basilar membrane). The elastic portion is held fixed along its boundaries and has the planform shape as shown in Figure 1. The system is driven by a prescribed fluid flux in the upper chamber at the left end of the tube (stapes).

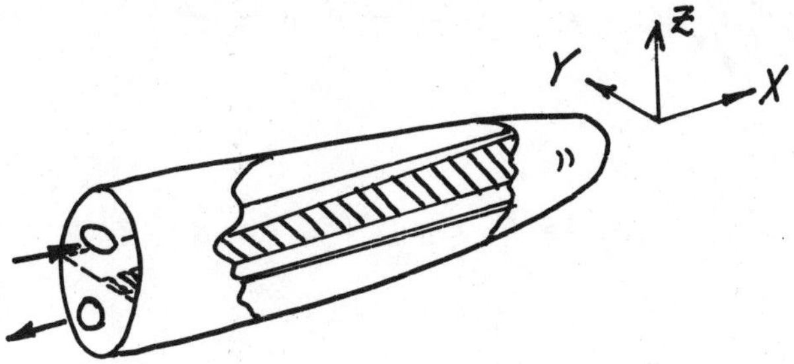

Fig. 7. Model hydroelastic system

The longitudinal coordinate X is directed along the centerline of the elastic portion of the dividing partition. The transverse coordinates Y, Z are such that Z is normal to the partition and Y is in its plane. The spatial coordinates and time are scaled by

$$x = X/L; \quad y = Y/B; \quad z = Z/B; \quad t = \Omega T$$

where L is the length, and B is a characteristic width of the partition. Ω is the frequency of the forcing function. The quantities of physical interest are the deflection of the basilar membrane W(X,Y,T), the pressure P(X,Y,Z,T) and the fluid velocity $\vec{U}$(X,Y,Z,T). The corresponding dimensionless quantities w, p, and $\vec{u}$ are defined by

$$W = W_A w(x,y,t)$$
$$P = P_A p(x,y,z,t)$$
$$\vec{U} = U_A \vec{u}(x,y,z,t)$$

with W_A, P_A, U_A being reference amplitudes. The dimensionless equations of the system are

$$\nabla_T \cdot \vec{u}_T + \epsilon \frac{\partial u_x}{\partial x} = 0 \tag{4-1}$$

$$\frac{\partial \vec{u}}{\partial t} = -\alpha \left\{ \nabla_T p + \epsilon \frac{\partial p}{\partial x} \vec{i} \right\} + \frac{1}{R} \left\{ \nabla_T^2 \vec{u} + \epsilon^2 \frac{\partial^2 \vec{u}}{\partial x^2} \right\} \tag{4-2}$$

$$\mathcal{D}(x) \frac{\partial^4 w}{\partial y^4} + \cdots + \mathcal{M}\, m(x) \frac{\partial^2 w}{\partial t^2} = -[p] \tag{4-3}$$

$$(u_x, u_y, u_z) = (0, 0, \partial w/\partial t) \text{ on } z = 0 \tag{4-4}$$

Other boundary conditions such as no flux through the rigid portion of the tube will be needed later. Eq. (4-1) is the continuity equation for an incompressible fluid. Eqs. (4-2) are the linearized momentum equations for a Newtonian incompressible fluid, sometimes called the (unsteady) Stokes equations. The quadratic convective acceleration term is neglected in (4-2) which is permissible for the assumed small motions. Eq. (4-3) is the equation of motion of the basilar membrane, and as written, expresses a balance between the net pressure acting on the membrane, its inertia, and a bending resistance due primarily to transverse curvature. The net pressure $[p] = p(x,y,o^+) - p(x,y,o^-)$. The implied neglected terms in (4-3) are terms involving the effect of longitudinal bending rigidity. Recent observations by Voldrich (1978) indicate that the vital basilar membrane is highly anisotropic in its elastic behavior, and has very little longitudinal coupling between transverse fibers. Eq. (4-3) expresses this fact and implies that to the first approximation it behaves like a continuous distribution of uncoupled transverse beams having variable stiffness and mass. Eq. (4-4) is also linearized for small displacements and is the usual no-slip and kinematic relation between velocity components and partition velocity. In these equations the spatial derivatives have been split into transverse and longitudinal operators where

$$\nabla_T = \vec{j}\frac{\partial}{\partial y} + \vec{k}\frac{\partial}{\partial z}$$

$$\nabla_T^2 = \frac{\partial^2}{\partial y^2} + \frac{\partial^2}{\partial z^2}$$

and $\vec{i}, \vec{j}, \vec{k}$ are unit vectors in the X,Y,Z directions. The dependent variables have been preliminarily scaled with

$$W_A = D_0/\rho B^4 \Omega^2; \quad P_A = \rho(\Omega W_A)^2; \quad V_A = \Omega W_A$$

where ρ is the fluid density, and D_o is a reference bending rigidity of a transverse beam

$$D_0 = E_0 h_0^3/12$$

where E_o is a reference Young's modules, and h_o is a reference plate thickness. This scaling leads to a system with the four parameters

$$\epsilon = B/L; \quad R = B^2\Omega/\nu; \quad m = \rho_p h_0/\rho W_A; \quad \alpha = W_A/B$$

where ν is the kinematic viscosity of the fluid (cm^2/sec), and ρ_p the density of the basilar membrane (gm/cm^3). ϵ is a geometrical slenderness parameter and is a small quantity. R is a Reynolds number with the interpretation that $1/\sqrt{R}$ is the ratio of the unsteady viscous boundary layer thickness $\sqrt{\nu/\Omega}$ to the characteristic transverse dimension B. $1/\sqrt{R}$ is a small quantity except at the very low end of the audible range, and decreases with increasing frequency. W_A can be interpreted as a reference wavelength and plays the same role as λ_0 in Section III. Thus α controls the dimensionality of the flow. At low frequency α is large, so the waves are long compared to the transverse dimension, and the flow is predominantly one-dimensional. At high frequency α is small, so the waves are short compared to the transverse dimension and the waves are of the "deep water" type. In the latter case the outer rigid boundary has a neglible effect on the flow. The mass parameter, m, is small at low frequency and increases with increasing frequency. It controls the importance of basilar membrane inertia, and can be interpreted as the ratio of basilar membane mass per unit area to the mass of fluid in a wavelength per unit area. The frequency dependence of the parameters is summarized is Fig. 8, where the following values have been used: B = 0.025 cm, L = 3.5 cm, ν = 0.01 cm^2/sec,

$D_o = 0.025$ dyne-cm, $\rho = \rho_p = 1$ gm/cm^3, and $h_o = 0.001$ cm. Also, in (4-3) D(x) and m(x) are O(1) smooth specified functions which give the longitudinal dependence of the transverse bending rigidity and mass of the basilar membrane.

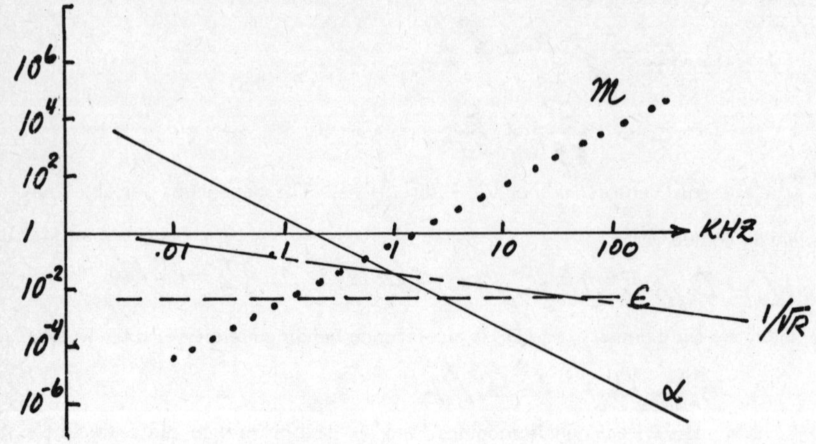

Fig. 8. Frequency dependence of parameters

Because the system (4-1) - (4-4) has four parameters which can take on a wide range of values as the frequency is changed, we are dealing with a complex situation with regard to asymptotic analysis. Many possibilities exist in terms of the type of expansions to use. The point of view taken here initially is that ϵ is the basic small parameter of the problem (independent of frequency) so that we can first of all try a WKB type of expansion as in Section III having the form

$$\begin{Bmatrix} \vec{u}(x,y,z,t) \\ p(x,y,z,t) \\ w(x,y,t) \end{Bmatrix} = e^{-i\tilde{x}} \begin{Bmatrix} \vec{q}_0(x,y,z,t) + \epsilon \vec{q}_1 + \cdots \\ \pi_0(x,y,z,t) + \epsilon \pi_1 + \cdots \\ \eta_0(x,y,t) + \epsilon \eta_1 + \cdots \end{Bmatrix} \tag{4-5}$$

$$\tilde{x}(x) = \epsilon^{-1} \int^{x} k(x)\,dx$$

where k(x) is phase function which must somehow be determined. These expansions when substituted into (4-1) - (4-4) given the dominant equations

$$\nabla_T \cdot \vec{q}_{0T} - ik q_{0x} = 0 \tag{4-6}$$

$$\frac{\partial \vec{q}_0}{\partial t} = -\alpha\{\nabla_T \Pi_0 - ik\Pi_0 \vec{i}\} + \frac{1}{R}\{\nabla_T^2 \vec{q}_0 - k^2 \vec{q}_0\} \tag{4-7}$$

$$\mathcal{D}(x)\frac{\partial^4 \eta_0}{\partial y^4} + \mathcal{M} m(x)\frac{\partial^2 \eta_0}{\partial t^2} = -[\Pi_0] = -2\Pi_0(x,y,0^+,t) \tag{4-8}$$

$$(q_{0x}, q_{0y}, q_{0z}) = (0, 0, \partial\eta_0/\partial t) \text{ on } z = 0 \tag{4-9}$$

An equation involving pressure only can be obtained by operating on (4-7) with $(\nabla_T - ik\vec{i})\cdot$ and using (4-6) to give

$$\nabla_T^2 \Pi_0 - k^2(x)\Pi_0 = 0 \tag{4-10}$$

Since there are no x-derivatives in this system, x plays the role of a parameter which is the major simplification of this method. Thus, the system (4-6)-(4-9) (plus additional boundary conditions yet to be specified) represents a homogeneous boundary value problem in the y,z cross plane. Degenerate versions of this type of problem were previously encountered in Sections II and III. The nature of this cross plane problem is that of an eigenvalue problem, with $\vec{q}_0$, Π_0, η_0 being eigenfunctions and k(x) the eigenvalue. The x-dependence of the eigenfunctions cannot be determined from the system (4-6)-(4-9). This information must come from consideration of the next higher order set of equations of the approximating sequence. In general k(x) depends on the shape of the cross section, boundary conditions, and the parameters α, R, $\mathcal{M}$. For extreme values of these parameters, as can occur depending on the frequency (cf. Fig. 8), various further asymptotic approximations can be considered for the cross plane problem. Some of these possibilities will be explored here.

Before discussing some properties of the cross plane problem in more detail, we can demonstrate for the inviscid limit $R \to \infty$ the conservation law which is the analogue of (3-18) in Section III,

$$k(x)\iint_{A(x)} \Pi_0^2(y,z;x)\,dy\,dz = \text{constant} \tag{4-11}$$

where $B^2A(x)$ is the physical cross sectional area of a chamber. Again, (4-11) has the physical interpretation that the rate of working (the integral of the product of pressure and longitudinal velocity) of the fluid is constant at each cross section. As in Section III, (4-11) can be used to find the x-variation of the first order eigenfunctions. To prove (4-11) consider the second order equations

$$\nabla_T^2\Pi_1 - k^2\Pi_1 = 2ik\frac{\partial \Pi_0}{\partial x} + ik'\Pi_0 \tag{4-12}$$

$$\frac{\partial \Pi_1}{\partial n} = \begin{cases} 0 & \text{on spiral ligament} \\ -\frac{1}{\alpha}\eta_1 & \text{on basilar membrane} \\ -[1+(\frac{\partial H}{\partial y})^2]^{\frac{1}{2}} ik\Pi_0\frac{\partial H}{\partial x} & \text{on } z = H(x,y) \end{cases} \tag{4-13}$$

$$\mathcal{D}(x)\frac{\partial^4\eta_1}{\partial y^4} - m_m(x)\eta_1 = -2\Pi_1(x,y,0^+) \tag{4-14}$$

where we are considering the simple harmonic motion case $\partial/\partial t \rightarrow i$. Here n is the outward normal to the cross plane boundary curve of the upper chamber, and $z = H(x,y)$ is surface of the rigid outer wall. Apply Green's formula,

$$\oint_{\partial A(x)} \left\{\Pi_1\frac{\partial \Pi_0}{\partial n} - \Pi_0\frac{\partial \Pi_1}{\partial n}\right\} ds = \iint_{A(x)} \{\Pi_1\nabla_T^2\Pi_0 - \Pi_0\nabla_T^2\Pi_1\}\, dA$$

substitute (4-10), (4-12), (4-13) and note that $ds = dy\{1 + (\partial H/\partial y)^2\}^{1/2}$ and $dA = H dy$ to obtain

$$\int_{-G(x)}^{G(x)} \Pi_1(x,y,0^+)\eta_0\,dy = \int_{-G(x)}^{G(x)} \Pi_0(x,y,0^+)\eta_1\,dy + \alpha\frac{\partial}{\partial x}ik\iint_{A(x)} \Pi_0^2\,dA \tag{4-15}$$

where the basilar membrane extends across $|y| \leq G(x)$. Now multiply (4-14) by η_0 and integrate across the width of the basilar membrane, substitute (4-15) and integrate the fourth derivative term by parts four times applying any combination of simply supported, clamped, or free boundary conditions on the beam to give

$$\int_{-G(x)}^{G(x)} \{ \mathcal{D}(x) \frac{\partial^4 \eta_0}{\partial y^4} - Mm(x)\eta_0 + 2\pi_0(x,y,\sigma) \} \eta_1 dy = -2\alpha \frac{\partial}{\partial x} ik \iint_{A(x)} \pi_0^2 dA$$

The integrand of the integral on the left hand side vanished by (4-8), from which (4-11) follows.

Before returning to the viscous problem it is instructive to consider in further detail a simplified inviscid problem having a cross plane geometry shown in Fig. 9. The domain is retangular with height H(x) and width 2G(x). On z=0 the basilar membrane spans the entire width, i.e. the spiral ligament has been omitted. Consider modes symmetric about y=0 (antisymmetric modes also exist and can be separately considered).

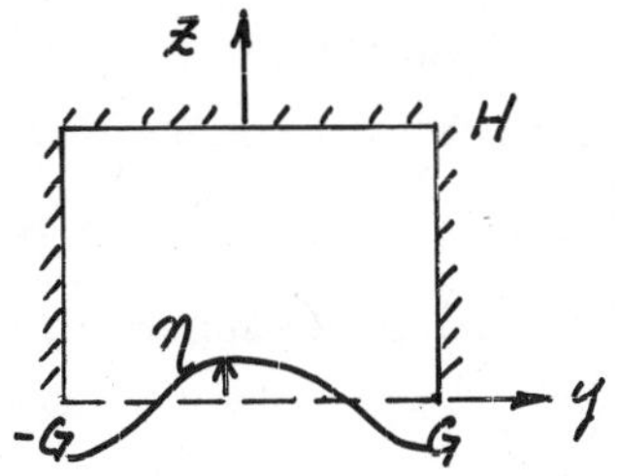

Fig. 9. Cross plan geometry of simplified problem.

Consider the forms

$$\pi_0 = F_n(z) \cos \gamma_n y$$
$$\eta_0 = \cos \gamma_n y \qquad (4\text{-}16)$$

where dependence on x as a parameter is understood. Here we allow the sides of the partition to slide up and down along the vertical walls, which enables us to obtain the simplest results. No flux through the side walls is satisfied if

$$\gamma_n = n\pi / G(x) \qquad n = 0, 1, 2, \ldots \qquad (4\text{-}17)$$

From (4-10) we obtain an equation for $F_n(z)$

$$F_n'' - (k_n^2 + \gamma_n^2) F_n = 0$$

which has the solution satisfying no flux through z=H and the kinematical condition on z=0

$$F_n = - \frac{\cosh\sqrt{k_n^2+\gamma_n^2}\,(z-H)}{\alpha\sqrt{k_n^2+\gamma_n^2}\,\sinh\sqrt{k_n^2+\gamma_n^2}\,H} \tag{4-18}$$

Substitution of (4-16) and (4-18) into (4-8) gives the dispersion relation

$$\mathcal{D}(x)\gamma_n^4 - \mathcal{M}m(x) = \frac{2H\coth\sqrt{k_n^2+\gamma_n^2}\,H}{\alpha\sqrt{k_n^2+\gamma_n^2}\,H} \tag{4-19}$$

Note that for n=0 there is no solution for real k, i.e. the n=0 mode does not propagate as a wave. However for $n \geq 1$ solutions exist for k real, but only in the range $x_t \leq x \leq x_r$. x_r is the "resonance" location where the left hand side vanishes and $k_n \rightarrow \infty$. x_t is a transition point which is that value of x which solves (4-19) with $k_n=0$. The x-dependence of the wavenumber as determined from (4-19) is shown in Fig. 10 for the case $\mathcal{D}$ =m=H=1, n=1, G(x)=(1+5x)/3, $\mathcal{M}\alpha \simeq h_o/B = 0.04$, α =1/400. Approximating the hyperbolic cotangent by unity leads to a curve which is indistinguishable from the exact solution.

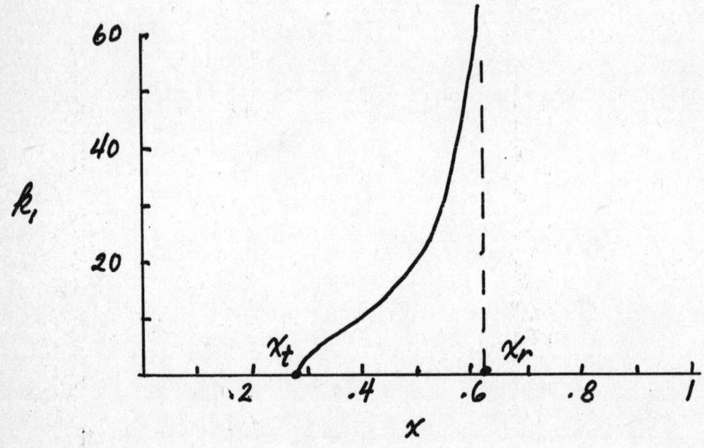

Fig. 10. Wavenumber dependence on longitudinal station as determined from Eq. (4-19).

Qualitatively the nature of the solution in the neighborhood of x_r is the same as that found near the resonance region of the two dimensional problem of Section III, ie. the wave is cut off for $x > x_r$. Including some form of partition damping, either the type used in Section III or a viscoelastic modulus making D(x) complex, would allow the present solution to be carried past the resonance region. A significant difference between the three dimensional and two dimensional problems considered here is the occurrence of a transition point in the former case. The WKB solution breaks down near x_t since the amplitude becomes infinite (cf. 4-11). A transition layer expansion can be constructed of the type used in Section I, and leads to a local Airy function solution which decays to the left of x_t. Thus the waveform has the nature of a packet decaying exponentially on both sides. This could provide a significant localization of the response, since $|x_r - x_t|$ can be made smaller than shown in Fig. 10, by using different parameters. Further work on this problem seems worthwhile.

Returning now to the viscous problem, we now consider the cross plane problem (4-6)-(4-9) at low frequency. In this limit α is large, m is small, and provided the frequency is not too small, R is still large (cf. Fig. 8). Since α is large we anticipate from (4-7) that $|\nabla_T \pi_0|$ is small, ie. the flow is predominately one dimensional. If this is the case it makes sense to first obtain the inviscid core flow and then correct for the viscous boundary layer on the surface. Consider the following expansions for the inviscid flow

$$\begin{aligned}
\pi_0 &= \pi_{00} + \frac{1}{\alpha}\pi_{01} + \cdots \\
\eta_0 &= \eta_{00} + \frac{1}{\alpha}\eta_{01} + \cdots \\
k &= \frac{1}{\sqrt{\alpha}} k_0 + \cdots \\
\vec{q}_{T0} &= \vec{q}_{T00} + \cdots \\
q_{0x} &= \sqrt{\alpha}\, q_{00x} + \cdots
\end{aligned} \qquad (4\text{-}20)$$

Then the dominant equations for the pressure are

$$\begin{aligned}
&\nabla_T^2 \pi_{00} = 0 \\
&\frac{\partial \pi_{00}}{\partial n} = 0 \quad \text{on boundary of upper chamber}
\end{aligned} \qquad (4\text{-}21)$$

which imply

$$\pi_{oo} = \pi_{oo}(x) \tag{4-22}$$

or the pressure is constant across each section. Since $\nabla_T \pi_{oo} = 0$, the flow has only a longitudinal component, and it also is uniform across the section. Of course the uniformity of the longitudinal velocity violates the no slip condition on the walls, so a viscous boundary layer is needed. Since the pressure does not vary across the basilar membrane the partition equation (4-8) becomes simply an equation for the static deflection of a transverse beam loaded by a uniform load.

$$D(x) \frac{\partial^4 \eta_{oo}}{\partial y^4} = -2 \pi_{oo}(x) \tag{4-23}$$

The solution of (4-23) for clamped edges at $y=\pm G(x)$, for example, is

$$\eta_{oo} = -\frac{1}{12} \pi_{oo}(x) \left\{ y^2 - G^2(x) \right\}^2 \tag{4-24}$$

We now correct the inviscid longitunal velocity $q_{oox} = k_o \pi_{oo}(x)$ for a boundary layer near a surface. Inside the boundary layer let

$$q_{ox} = \sqrt{\alpha} \, Q_{oox}(x,\zeta) + \cdots \tag{4-25}$$

where ζ is a boundary layer coordinate measureing the stretched distance from a surface

$$\zeta = -n \sqrt{R} \tag{4-26}$$

where n here is an unstretched coordinate measuring distance along an outward normal. ζ is 0(1) when n is $0(1/\sqrt{R})$, the boundary layer thickness. Using (4-25), and (4-26) the x-component of (4-7) becomes

$$\frac{\partial^2 Q_{oox}}{\partial \zeta^2} - i \, Q_{oox} = -i k_o \pi_{oo}(x) \tag{4-27}$$

which has the solution

$$Q_{oox} = q_{oox}(x)\left(1 - e^{-\sqrt{i}\,\zeta}\right) \tag{4-28}$$

Note that Q_{oox} vanishes on the wall, $\zeta = 0$, and approaches the inviscid velocity q_{oox} as $\zeta \to \infty$. Thus (4-28) is valid over the entire cross section.

The $0(1/\alpha)$ equations for the pressure are

$$\nabla_T^2 \Pi_{01} = k_0^2 \Pi_{00}$$
$$\frac{\partial \Pi_{01}}{\partial n} = \begin{cases} -\eta_{00} & \text{on partition} \\ 0 & \text{on other boundaries} \end{cases} \tag{4-29}$$

Applying the divergence theorem

$$\iint_{A(x)} \nabla_T^2 \Pi_{01}\, dA = \oint_{\partial A(x)} \frac{\partial \Pi_{01}}{\partial n}\, ds$$

we obtain

$$\iint_{A(x)} k_0^2 \Pi_{00}\, dA = -\int_{-G(x)}^{G(x)} \eta_{00}\, dy \tag{4-30}$$

which has the interpretation of a one dimensional continuity or mass conservation equation. Since $k_o^2 \Pi_{oo} = k_o q_{oox}$ we can replace q_{oox} with the uniformly valid Q_{oox}. Then evaluating the integrals using (4-28) and (4-24) we find the relation for $k_o(x)$

$$k_0^2(x) = \frac{4}{45}\, \frac{G^5(x)}{\lambda \mathcal{Q}(x) A^*(x)} \tag{4-31}$$

where $A^*(x)$ is the reduced area

$$A^*(x) = A(x) - \frac{\mathcal{P}(x)}{\sqrt{iR}} \tag{4-32}$$

where $\mathcal{P}(x)$ is the dimensionless perimeter of the cross section. To determine $\Pi_{oo}(x)$ we use conservation of rate of working, (4-11), but integrate over $A^*(x)$ instead of $A(x)$ to correct for the flux deficit in the boundary layer, with the result

$$\Pi_{\infty}(x) = \frac{C_0}{\sqrt{k_0(x)A^*(x)}} \tag{4-33}$$

where C_0 is an arbitrary constant. The dominant solution for the pressure can be written

$$p \sim C_0 \frac{\mathcal{D}(x)^{1/4}}{A^*(x)^{1/4} G(x)^{5/4}} \left\{ \frac{e^{\frac{i}{\epsilon\sqrt{\alpha}}(\xi-\xi_1+t)} e^{\frac{1}{\epsilon\sqrt{\alpha}}(\gamma-\gamma_1)} - e^{-\frac{i}{\epsilon\sqrt{\alpha}}(\xi-\xi_1-t)} e^{-\frac{1}{\epsilon\sqrt{\alpha}}(\gamma-\gamma_1)}}{2\cos\left(\frac{1}{\epsilon\sqrt{\alpha}}\int_0^1 k_0(x)dx\right.} \right\}$$

(4-34)

with

$$\xi(x) = \frac{2}{\sqrt{45}} \int_0^x \sqrt{\frac{G^5(x)}{A(x)\mathcal{D}(x)}}\, dx \; ; \quad \xi_1 = \xi(1)$$

$$\gamma(x) = \frac{1}{\sqrt{90R}} \int_0^x \mathcal{P}(x) \sqrt{\frac{G^5(x)}{A^3(x)\mathcal{D}(x)}}\, dx \; ; \; \gamma_1 = \gamma(1) \tag{4-35}$$

The form (4-34) satisfies pressure equalization at the helicotrema, x=1, and the constant C_0 can be determined from a specified flux at the stapes, x=0.

A different approach to the low frequency behavior was taken by Chadwick and Cole (1979) and Holmes (1980) who showed that a direct long wave expansion of the full problem leads to an ordinary differential equation for the pressure

$$\frac{d}{dx}\left(A^* \frac{dp}{dx}\right) + \frac{4}{45} \frac{1}{\epsilon^2\alpha} \frac{G^5(x)}{\mathcal{D}(x)} p(x) = 0 \tag{4-36}$$

What is interesting in the present context is that the first term of the asymptotic solution to (4-36) for small $\epsilon^2\alpha$ leads to a result identical with the theory presented here. Comparison of predictions from the low frequency theory with experiments on a mechanical model are good and can be found in Chadwick et.al (1980).

We now turn to some discussion of the high frequency behavior. Referring to Fig. 8 we can consider the limit $\alpha \to 0$, $R \to \infty$, and $\mathcal{M} = O(1)$. With $\alpha \to 0$ in (4-7) we can anticipate that $|\nabla_T \Pi_0|$ is large. This could be accomplished by either making the pressure itself large, or by stretching the vertical coordinate. The former possibility leads to an inconsistency since a large pressure implies a large deflection by (4-8) which in turn implies a large vertical velocity by (4-9) which fails to balance the vertical presure gradient in (4-7). Thus we must stretch the vertical coordinate at the partition. This will lead to an inviscid boundary layer whereby the dominant fluid motion is confined to a thin layer (thin with respect to the transverse dimension) near the partition. This is in essence the "deep water" approximation, and the outer wall should have neglible influence. At the same time we can expect a viscous boundary layer needed to satisfy the no slip condition. The characteristic feature of the high frequency behavior is the existance of two boundary layers and their mutual interation. The degree of interaction depends on the relative thickness of the two boundary layers. Here we shall consider the case of full interaction which occurs when both boundary layers have the same thickness. This is the case when $\alpha \rightarrow 0$ and $R \rightarrow \infty$ in such a way that $\alpha \sqrt{R} = 0(1)$. A sketch of the anticipated length scales relative to the characteristic cross plane dimension is shown in Fig. 11.

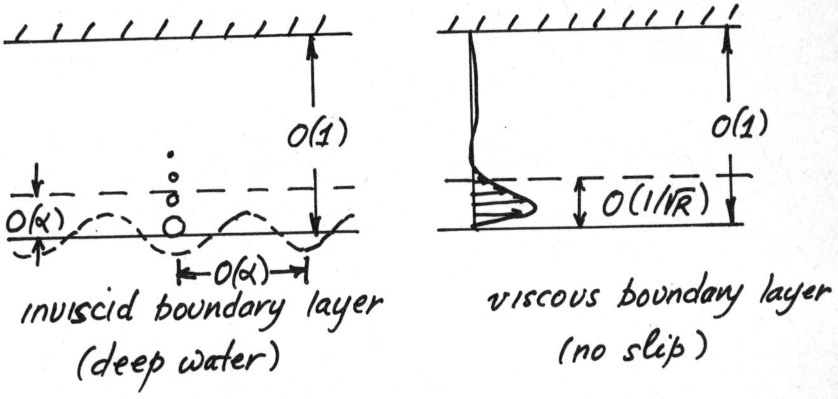

Fig. 11. Boundary layer length scales

Instead of using boundary layer type expansions on the cross plane problem (4-6) -(4-9) it is somewhat more efficient to try a combined WKB-boundary layer expansion on the original system (4-1)-(4-4). Consider then

$$\left\{\begin{array}{l}\vec{u}(x,y,z,t)\\ p(x,y,z,t)\\ w(x,y,t)\end{array}\right\} = e^{-i\tilde{x}}\left\{\begin{array}{l}\vec{q}_0(x,y,z^*,t)+\frac{\epsilon}{\sqrt{R}}\vec{q}_1+\cdots\\ \pi_0(x,y,z^*,t)+\frac{\epsilon}{\sqrt{R}}\pi_1+\cdots\\ \eta_0(x,y,t)+\frac{\epsilon}{\sqrt{R}}\eta_1+\cdots\end{array}\right.$$

$$\tilde{x}(x) = \frac{\sqrt{R}}{\epsilon}\int^{x} k(x)dx\ ; \quad z^* = \sqrt{R}\, z \tag{4-37}$$

In this limit the wavelength is the same order as the boundary layer thickness. This can evidently happen only with no longitudinal coupling of adjacent transverse beam elements. Eqs. (4-37) then lead to the following dominant equations for the simple harmonic motion case

$$\frac{\partial q_{0z}}{\partial z^*} - ik\, q_{0x} = 0 \tag{4-38}$$

$$\left\{\frac{\partial^2}{\partial z^{*2}} - (k^2+i)\right\}\vec{q}_0 = \alpha\sqrt{R}\left(-ik\pi_0\vec{i} - \frac{1}{\epsilon\sqrt{R}}\frac{\partial \pi_0}{\partial y}\vec{j} + \frac{\partial \pi_0}{\partial z^*}\vec{k}\right) \tag{4-39}$$

$$D(x)\frac{\partial^4 \eta_0}{\partial y^4} - \mathcal{M}m(x)\eta_0 = -2\pi_0(x,y,0^+) \tag{4-40}$$

$$(q_{0x}, q_{0y}, q_{0z}) = (0,0,i\eta_0) \text{ on } z^*=0 \tag{4-41}$$

$$\text{decay as } z^*\to\infty$$

$$\frac{\partial^2 \pi_0}{\partial z^{*2}} - k^2\pi_0 = 0 \tag{4-42}$$

The system can be simply solved by first taking

$$\Pi_0(x,y,z^*) = \Pi_0(x,y,0)\, e^{-kz^*} \qquad (4\text{-}43)$$

as the appropriate solution of (4-42), which implies that k should have a positive real part to insure decay away from the partition. Then solutions to (4-39) can be obtained with unknown constants appearing in the homogeneous solutions. These constants can be uniquely determined from (4-28) and the boundary conditions. In particular, we find

$$\Pi_0(x,y,0) = \frac{\eta_0(x,y)}{\alpha\sqrt{R}}\,\frac{k^*}{k(k-k^*)} \qquad (4\text{-}44)$$

$$k^* = \sqrt{k^2+i}$$

which gives the equation for η_0

$$\mathcal{D}(x)\frac{\partial^4\eta_0}{\partial y^4} - \left\{\mathcal{M}\, m(x) + \frac{2}{\alpha\sqrt{R}}\,\frac{k^*}{k(k-k^*)}\right\}\eta_0 = 0 \qquad (4\text{-}45)$$

Eq. (4-45) together with boundary conditions on $y=\pm G(x)$ represents the free vibrations of a beam with a complex virtual mass. The eigenvalue k is in the virtual mass term. For fixed x, there exists a discrete set of transverse mode eigenfunctins η_{o_n} and corresponding complex eigenvalues k_n, n=1, 2, ... Problems of this sort for different boundary conditions have been considered in detail by Rayleigh (1945). The solutions of (4-45) are linear combinations of trigonometric and hyperbolic functions. Satisfying the boundary conditions then leads to the equation determining the k_n

$$\frac{k_n^*}{k_n(k_n^*-k_n)} = \frac{1}{2}\sqrt{R}\,\alpha\left\{\mathcal{D}(x)\left(\frac{\beta_n}{2G(x)}\right)^4 - \mathcal{M}\, m(x)\right\} \qquad (4\text{-}46)$$

The β_n depend on the boundary conditions, eg.

$$\beta_1 = \begin{cases} \pi & \text{simply supported} \\ 4.73 & \text{clamped} \end{cases}$$

Denoting the right hand side of (4-46) by f(x) we summarize the first order solutions

$$\pi_0(x,y,z^*) = -\frac{f(x)}{\alpha\sqrt{R}}\,\eta_0(x,y)\,e^{-kz^*}$$

$$g_{0x}(x,y,z^*) = -k(x)f(x)\,\eta_0(x,y)\left\{e^{-kz^*} - e^{-k^*z^*}\right\} \qquad (4\text{-}47)$$

$$g_{0y}(x,y,z^*) = -\frac{if(x)}{\epsilon\sqrt{R}}\,\frac{\partial\eta_0(x,y)}{\partial y}\left\{e^{-kz^*} - e^{-k^*z^*}\right\}$$

$$g_{0z}(x,y,z^*) = i\,k(x)f(x)\,\eta_0(x,y)\left\{e^{-kz^*} - \frac{k}{k^*}e^{-k^*z^*}\right\}$$

where the n subscripts are understood. The root locus of the root computed from (4-46) is shown in Fig. 12, where only the fourth quadrant root (representing a damped right running wave) is given

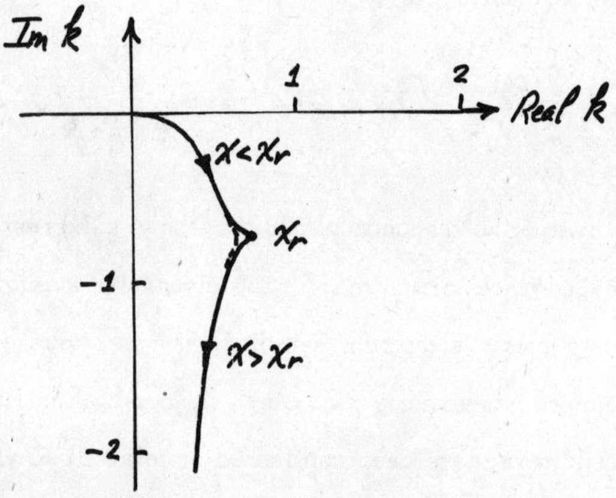

Fig. 12. Root locus plot of root corresponding to right running wave computed from Eq. (4-46)

The direction of increasing x is shown by the arrows. It is seen from (4-46) that when the right hand side vanishes, $k^*=0$, or $k=\sqrt{-i}$. This point is labeled r in Fig. 12. This point defines a "resonance" location x_r, and the maximal response can be expected to occur near this location. x_r is frequency dependent and decreases monotonically as the frequency increases. In a neighborhood of x_r the boundary layer theory locally breaks down since $k^*\to 0$ and the exponentials in (4-47) don't decay. In particular, the vertical component of velocity, $q_{oz} \sim \eta_o(x,y)$, which has the interpretation of a narrow vertical jet flow near resonance. Another indication of a local breakdown in the theory is that the amplitude functions become singular at the resonance location. Going through a rather tedious solution of the second order equations which is too long to repeat here, we find that near resonance the amplitude is proportional to $(k^*)^{-1/2}$. The author is currently studying physical mechanisms which will eliminate the singularity. One such mechanism is to include a small longitudinal coupling term in the partition equation (43) which has the effect of rounding off the pointed root trajectory near the resonance location as shown by the dotted line in Fig. 12, and prevents k* from being zero.

Acknowledgements

Part of this research was carried out with support of NSF under Grant ENG 76-81574 while the author was at the University of California, Los Angeles, Dept. of Mechanics and Structures. The author is grateful for many discussions with Professors J. D. Cole, M. E. Fourney, and L. P. Cook.

REFERENCES

1. Allen, J.B. (1977). Two-dimensional cochlear fluid model: New results, J. Acoust. Soc. Am. 61, 110-119.

2. Allen, J.B. and Sondhi, M.M. (1979). Cochlear macromechanics: Time domain solutions, J. Acoust. Soc. Am. 66, 123-132.

3. Boer, E. de (1979). Short-wave world revisited: Resonance in a two-dimensional cochlear model, Hearing Res. 1, 253-281.

4. Chadwick, R.S. (1978). Vibrations of long narrow plates-II, Quart. Applied Math., XXXVI, no. 2, 155-166.

5. Chadwick, R.S. and Cole, J.D. (1979). Modes and waves in the cochlea. Mechanics Research Communications, 6(3), 177-184.

6. Chadwick, R.S., Fourney, M.E. and Neiswander, P. (1980). Modes and waves in a cochlear model. Hearing Res. 2, 475-483.

7. Cole, J.D. (1968) Peturbation Methods in Applied Mathematics. Blaisdell Publishing Co., Waltham, Mass.

8. Goodier, J.N. and Fuller, F.B. (1964). Turning-point peculiarities in the near resonant resonse of a slightly tapered membrane setup, J. Acoust. Soc. Am., 46, no. 8, 1491-1495.

9. Holmes, M. (1980). Low frequency asymptotics for a hydroelastic model of the cochlea. SIAM J. Appl. Math 38(3), 445-456.

10. Lesser, M.B. and Berkley, D.A. (1972). Fluid mechanics of the cochlea. Part 1, J. Fluid Mech. 51, 497-512.

11. Lighthill, M.J. (1980) Energy flow in the cochlea, to appear in J. Fluid Mech.

12. Neeley, S.T. (1980). A two-dimensional mathematical model of the mechanics of the cochlea, submitted to J. Acoust. Soc. Am.

13. Ranke, O.F. (1950). Theory of operation of the cochlea: A contribution to the hydrodynamics of the cochlea, J. Acoust. Soc. Am. 22, 772-777.

14. Rayleigh, J.W.S. (1945). Theory of Sound, vol. I. Dover Publications, New York.

15. Siebert, W.M. (1974). Ranke revisited - a simple short wave cochlea model, J. Acoust.Soc. Am. 56, 594-600.

16. Sondhi, M.M. (1978). Method for computing motion in a two dimensional cochlear model, J. Acoust. Soc. Am. 63, 1468-1477.

17. Steele, C.R. (1981). Lecture notes on cochlear mechanics. NSF-CBMS regional conference on Mathematical Modeling of the Hearing Process, Rensselaer Polytechnic Institute, Troy, N.Y., also to appear in SIAM Regional conf. Series on Appl. Math.

18. Steele, C.R. and Taber, L.A. (1979). Comparison of WKB and finite differences calculations for a two dimensional cochlear model, J. Acoust. Soc. Am. 65, 1001-1006.

19. Viergever, M.A. (1980). Mechanics of the Inner Ear. Delft University Press, The Netherlands.

20. Voldrich, L. (1978). Mecchanical properties of the basilar membrane. Acta Oto larygl. 86, 331-335.

A HYDROELASTIC MODEL OF THE COCHLEA: AN ANALYSIS FOR LOW FREQUENCIES

Mark H. Holmes
Department of Mathematical Sciences
Rensselaer Polytechnic Institute
Troy, New York 12181

Introduction

A number of mathematical models of the low frequency behavior of the cochlea have appeared over the years, beginning with Helmholtz (1885), and continuing with Peterson and Bogert (1950) and Zwislocki (1965). This article deals, principally, with the theory developed by the author in conjunction with Dr. Richard Chadwick and Dr. Julian Cole. It differs from the earlier "long wave" theories in that it is based on modeling the cochlea using continuum theory and then reducing, when possible, the problem through the use of asymptotic approximations. In doing this, some of the results are the same as found in the earlier theories. However, there are also differences, relating to the contribution of the basilar membrane and fluid viscosity. Also, with our approach, the extension to more complicated geometries and higher frequency regimes is possible (e.g., Holmes, 1982).

Formulation and Scaling of Model

In what follows, the cochlea is idealized to consist of an unrolled tapered tube containing two chambers that are each filled with an incompressible viscous fluid. The chambers are separated by a planar region which contains a rigid section, that represents the bony shelf, and a flexible portion, Γ, that represents the basilar membrane. At the apical end the two chambers are connected by an aperature in the partition, known as the helicotrema. The outer boundary of the cochlea consists of a rigid portion, called the cochlear wall, and two openings at the basal end that are each covered by a flexible membrane. The stapes transmits the signals from the outer ear to the cochlea by pushing against the upper opening Γ_W, which is known as the oval window. The lower opening, the round window, is represented by Γ_R. For simplicity, the cochlear wall is assumed to be symmetric through the x,y-plane. Also, away from the ends $x = 0, 1$ it is assumed that the boundary of the basilar membrane can be written as

$$y = \pm G(x) \ .$$

The part of the boundary given by $y = G(x)$ represents the portion attached to the spiral lamina, and $y = -G(x)$ is the portion attached to the spiral ligament.

The basilar membrane is modeled as a linear orthotropic elastic plate with a uniform thickness, and which is simply supported along its boundary. The deflection of the plate is represented by $\eta(x,y,t)$. The velocity and pressure of the fluid are represented by $\vec{v}(x,y,z,t)$ and $p(x,y,z,t)$, respectively. In what follows, both the dependent and independent variables are assumed to be dimensionless. Asterisks are used to indicate their dimensional analogs. So, for example, the nondimensional spatial coordinates x,y,z are related to x^*,y^*,z^* as follows

$$x^* = Lx, \qquad y^* = By, \qquad z^* = Bz,$$

where L and B are the length and width of the basilar membrane, respectively.

In nondimensional form, the equations describing the motion are (Holmes, 1981):

(i) for the fluid

$$\frac{\partial}{\partial t} \vec{v} - \delta^2 \left(\varepsilon^2 \frac{\partial^2}{\partial x^2} + \frac{\partial^2}{\partial y^2} + \frac{\partial^2}{\partial z^2}\right) \vec{v} = -\nabla p \ , \tag{1a}$$

$$\varepsilon^2 \frac{\partial}{\partial x} v_1 + \frac{\partial}{\partial y} v_2 + \frac{\partial}{\partial z} v_3 = 0 \tag{1b}$$

(ii) for the basilar membrane

$$\frac{\partial^4}{\partial y^4} \eta + 2\varepsilon^2 D_3 \frac{\partial^4}{\partial x^2 \partial y^2} \eta + \varepsilon^4 D_1 \frac{\partial^4}{\partial x^4} \eta + \frac{\varepsilon^2}{\alpha} \frac{\partial^2}{\partial t^2} \eta \tag{1c}$$

$$= - [\![p(x,y,0,t)]\!]_\Gamma \ ,$$

The parameters are

$$\varepsilon = \frac{B}{L} \ , \qquad \delta^2 = \frac{\nu}{B^2 \bar{\omega}_0} \ , \qquad \alpha = \frac{\rho B}{\mu} \ ,$$

where ρ and ν represent the density and kinematic viscosity of the fluid, and μ is the mass density of the plate. Also,

$$\hat{\omega}_0^2 = \frac{D_2^*}{\rho L^2 B^3} \ ,$$

where D_2^* is the (dimensional) bending rigidity of the plate in the y direction, and

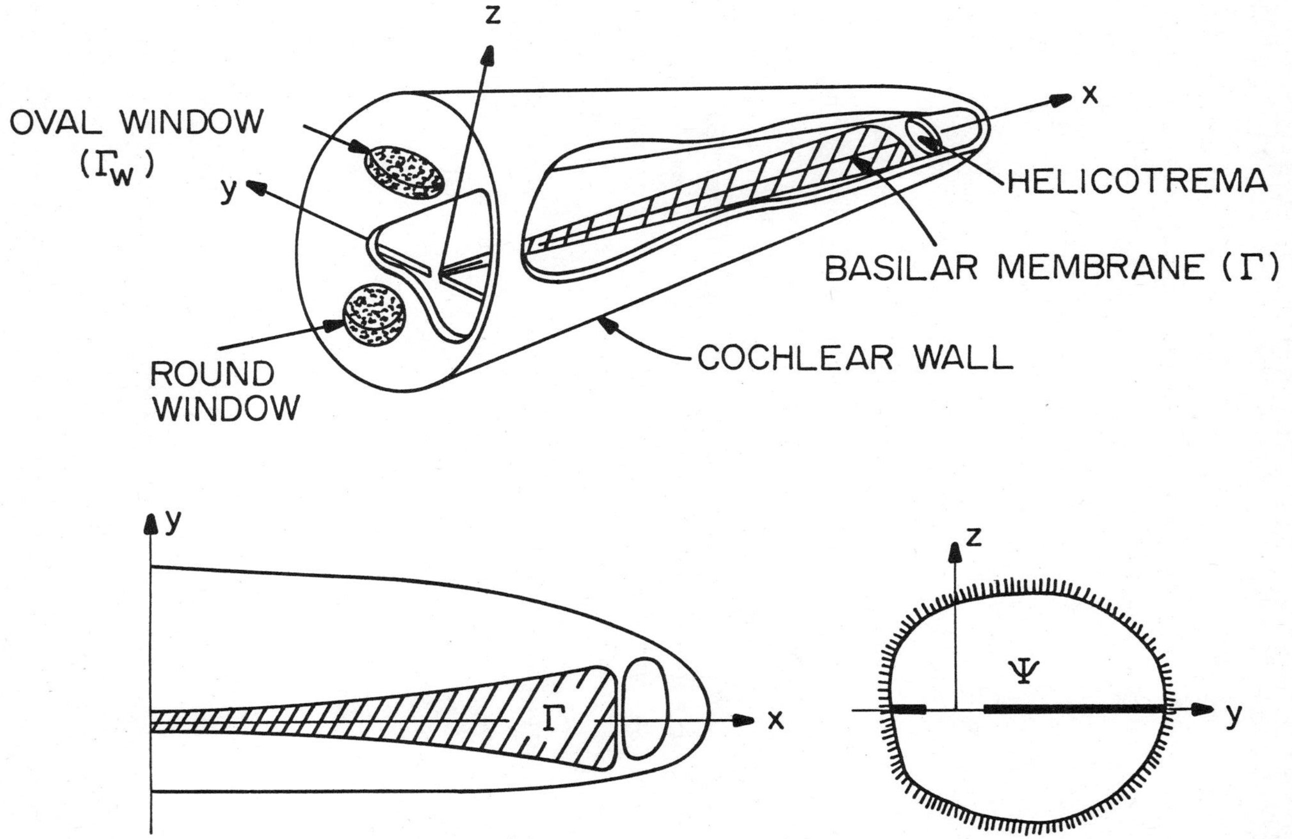

Fig. Geometry for the hydroelastic model of the cochlea. Shown are a cutaway of the entire cochlea, the plan form of the partition, and the cross section of the cochlea.

$$[\![p]\!]_\Gamma = p(x,y,0^+,t) - p(x,y,0^-,t) \ .$$

The D_1 and D_3 in (1c) represent the respective bending and twisting rigidities normalized by D_2^*.

The fluid velocity satisfies the usual kinematic condition on the basilar membrane and the two windows. In particular,

$$\vec{v} = \begin{cases} (0,0,\varepsilon^2 \frac{\partial}{\partial t} \eta) & \text{on } \Gamma \\ (\frac{\partial}{\partial t} \eta_w, 0, 0) & \text{on } \Gamma_w \\ -(\frac{\partial}{\partial t} \eta_w, 0, 0) & \text{on } \Gamma_R \ . \end{cases}$$

In the last two boundary conditions η_w is a known function and represents the driving of the oval window by the stapes. The boundary condition on Γ_R is a reflection of both the symmetry of the geometry and the conservation of the fluid volume. Finally, note that $\vec{v}$ satisfies the no-slip condition on the rigid portion of the cochlear wall and the bony shelf. As a final comment on the notation, the cross section of the upper chamber is represented by $\Psi(x)$, and its boundary is denoted by $\partial\Psi$.

Slender Body Approximation

The problem can be simplified by taking advantage of the fact that ε is small for most mammalian cochlae. For example, for humans, $\varepsilon^2 \sim 10^{-3}$. The appropriate asymptotic expansions in this case are

$$\vec{v} \sim (v_o,0,0) + \varepsilon\vec{v}_1 + \ldots \ ,$$

$$p \sim p_o(x,y,z,t) + \varepsilon p_1 + \ldots \ ,$$

and

$$\eta \sim \eta_o(x,y,t) + \varepsilon\eta_1 + \ldots \ .$$

Introducing these into (1) one finds from the O(1) problem that

$$p_o^{\pm} = p_o^{\pm}(x,t) \ , \tag{2}$$

$$\frac{\partial}{\partial t} v_o^{\pm} - \delta^2 \nabla^2_{y,z} v_o^{\pm} = - \frac{\partial}{\partial x} p_o^{\pm} \ , \tag{3}$$

and

$$\eta_o = \zeta_o(x,y) [\![p_o(x,t)]\!] \ , \tag{4}$$

where

$$\zeta_0(x,y) = -\frac{1}{24}\,[y^2 - G^2(x)]\;[y^2 - 5G^2(x)] \quad . \tag{5}$$

The +, - refer to the upper and lower chambers, respectively. Also,

$$\nabla^2_{y,z} = \frac{\partial^2}{\partial y^2} + \frac{\partial^2}{\partial z^2} \quad .$$

The equation for $p_0^{\pm}$ is obtained by applying Green's theorem to the $O(\varepsilon^2)$ continuity equation. This solvability condition results in the following reduced system of equations

$$\frac{\partial}{\partial x}[A(x)\frac{\partial}{\partial x}p_0 - \delta^2\int_{\partial\Psi}\frac{\partial}{\partial n}v_0] = K(x)\frac{\partial^2}{\partial t^2}p_0 \ , \tag{6a}$$

and

$$\frac{\partial}{\partial t}v_0 - \delta^2\nabla^2_{y,z}v_0 = -\frac{\partial}{\partial x}p_0 \quad \text{in } \Psi(x) \quad , \tag{6b}$$

where

$$p_0 = \tfrac{1}{2}\,[\![p_0(x,t)]\!]_\Gamma \quad ,$$

$$v_0 = \tfrac{1}{2}(v_0^+ - v_0^-) \quad ,$$

$$K(x)\;\; k_0^2\; G^5(x) \quad ,$$

and $k_0^2 = \frac{8}{15}$. The function $A(x)$ in (6a) represents the area of $\Psi(x)$, and n is the unit outward normal to $\partial\Psi$ in the y,z - plane .

Consequently, to the first order, there is unsteady Poiseuille in both chambers and the plate reduces to a massless beam in the transverse direction. The contribution of the viscosity is through the shear stress integrated over the cross section. As for the basilar membrane, its contribution in the reduced problem is contained entirely in the coefficient $K(x)$. So, for example, changing the boundary conditions on the plate only effects the constant k_0. In this sense, stiffening the boundary conditions, such as changing to a clamped plate, increases $K(x)$. Along different lines, note that if the thickness of the plate is assumed to depend on the longitudinal variable x then

$$K(x) = k_0^2\,\frac{G^5(x)}{D(x)} \quad , \tag{7}$$

where $D(x)$ represents the normalized bending rigidity in the transverse direction (Holmes, 1980a). Therefore, in this theory, $\frac{1}{2}K(x)$ represents the (normalized) volume compliance per unit length of the partition.

To complete the problem, the boundary conditions for $p_o(x,t)$ need to be specified. At the apical end, because of the helicotrema, we take

$$p_o(1,t) = 0 \ . \tag{6c}$$

The condition at the basal end comes from the conservation of fluid volume. It is found that this leads to the following condition when $x = 0$

$$A_o \frac{\partial}{\partial x} p_o - \delta^2 \int_{\partial\Psi} \frac{\partial}{\partial n} v_o = - \frac{d^2}{dt^2} \int_{\Gamma_w} \eta_w \ , \tag{6d}$$

where $A_o = A(0)$. Finally, note that from the no-slip condition

$$v_o \Big|_{\partial\Psi} = 0 \ . \tag{6e}$$

Viscous Boundary Layer

The problem for the pressure is now complete. However, even for a very simple geometry it is difficult to solve it in closed form. Accordingly, to be able at least to determine the qualitative behavior of the solution, we now consider expansions in terms of δ. This is appropriate as even at the lower threshold of hearing, which is approximately 20 Hz, the boundary-layer thickness is less than 10^{-1}.

The use of boundary-layer theory to find v_o from (6b,e) involves the introduction of orthogonal coordinates, corresponding to the normal and tangential directions along $\partial\Psi(x)$, and the consideration of a boundary-layer in the normal direction. Carrying out the details one finds that, to the first order in δ,

$$\delta^2 \int_{\partial\Psi} \frac{\partial}{\partial n} v_o = \frac{\delta P(x)}{\sqrt{\pi}} \frac{\partial}{\partial x} \int_0^t \frac{p_o(x,\tau)}{\sqrt{t-\tau}} d\tau \ , \tag{8}$$

where $P(x)$ represents the perimeter of $\Psi(x)$.

In what follows we are concerned, principally, with the response for a periodic forcing of the oval window. In this case, the longtime solution of (6b,e) gives the following

$$\delta^2 \int_{\partial\Psi} \frac{\partial}{\partial n} v_o = \frac{\delta P(x)}{\sqrt{i\omega}} \frac{\partial}{\partial x} p_o \ , \tag{9}$$

where ω is the driving frequency. From (9) we can be more precise in specifying the domain of validity of the boundary layer approximation. In particular, it is required that

$$\frac{\delta}{2r(x)\sqrt{\omega}} << 1 \quad , \tag{10}$$

where r(x) is the "hydraulic radius" of $\Psi(x)$ and is given as

$$r(x) = \frac{A(x)}{P(x)} \quad .$$

In the human cochlea, (10) is satisfied throughout the normal range of hearing.

Solution of Reduced Initial Value Problem

We now solve the reduced problem (6a) for the pressure jump by taking advantage of the hyperbolic character of the equation. This assumes that the damping is small, which is consistent with the boundary layer expansions used in the last section. An extended discussion of the analysis presented here can be found in Holmes (1981), although, the geometry is somewhat more restricted than that used here.

To begin, (6a) is solved for a semi-infinite geometry. So, the boundary condition at the basal end is to be satisfied and it is assumed that no energy is coming in from infinity. Also, it is assumed that the coefficients are slowly varying functions of x. To expand p_o in terms of δ it is advantageous to introduce a time variable ζ given as

$$\zeta = t - \int_0^x k(\xi)d\xi \quad ,$$

where

$$k(x) = \sqrt{\frac{K(x)}{A(x)}} \quad .$$

This choice for ζ comes from the characteristic for the inviscid problem. Using a multivariable expansion in δ one finds that

$$p_o(x,t) = \int_0^{t-c(x)} \mathcal{F}(x,t-c-\tau)\eta_w''(\tau)d\tau \quad , \tag{11}$$

where

$$\mathcal{F}(x,t) = \frac{A_w}{A_o k(x)} \operatorname{erfc}\left[\frac{\delta g(x)}{2\sqrt{t}}\right] \quad ,$$

$$g(x) = \int_0^x \frac{k(\xi)}{r(\xi)}d\xi$$

and

$$c(x) = \int_0^x k(\xi)d\xi \quad .$$

Also, it is assumed here that the forcing function η_w in (6d) is independent of the spatial coordinates, and so, A_w represents the normalized area of the oval window.

It is understood in (11) that $p_o = 0$ for $t \leq c(x)$. Consequently, (11) gives the exact solution of (5a) so long as

$$t \leq \int_0^1 k(\xi)d\xi \quad ,$$

which is the time it takes the disturbance to reach the apical end. For t larger than c(1) it is necessary to include in (11) the reflection of the wave from the apical end. In this way, the general solution consists of the superposition of waves, traveling to the left and right. However, because of the viscous decay, the wave that dominates at any particular instant is the one that has not yet reflected at the apical end. Accordingly, for a periodic forcing, at a moderately high frequency, the solution of the initial value problem evolves into a wave that travels towards the helicotrema. Very near the helicotrema, however, the response takes on the appearance of a standing wave because of the similarity of the wave with its reflection in this region. The extent of this region depends on the frequency, and it decreases as the frequency increases.

Longtime Problem

We now investigate the longtime behavior of the system for the case of a periodic forcing of the oval window. With this in mind it is assumed that

$$\eta_w = e^{i\omega t} \quad ,$$

and

$$p_o(x,t) = p_L(x)e^{i\omega t} \quad .$$

Using (9) the problem reduces to a Sturm-Liouville problem for $p_L(x)$. In particular, one obtains the following

$$\frac{d}{dx}[H(x)\frac{d}{dx}p_L] = -\omega^2 K(x)p_L \quad , \tag{12a}$$

where

$$H(0)\frac{d}{dx}p_L(0) = \omega^2 A_w \quad , \tag{12b}$$

$$p_L(1) = 0 \quad , \tag{12c}$$

and

$$H(x) = A(x) - \frac{\delta}{\sqrt{i\omega}} P(x) \quad . \tag{13}$$

The above problem for $p_L(x)$ is simple enough that it can be solved exactly for reasonably realistic geometries (Holmes, 1980b). It can also be solved, approximately, using slowly varying theory and a WKB type expansion, as in Chadwick and Cole (1979) and Holmes (1980b). This latter approach is related to the expansions used in analyzing the transient problem, and so, it is the one that will be used for the longtime problem.

The appropriate expansion for $p_L(x)$ is

$$p_L(x) \sim \omega[a_o(x) + \ldots]e^{i\omega\theta + i\sqrt{\omega}\psi} \tag{14}$$

Substituting this into (12a) one finds from the $O(\omega^2)$ problem that

$$\left(\frac{d}{dx}\theta\right)^2 = \frac{K(x)}{A(x)} \quad ,$$

and from the $O(\omega^{3/2})$ problem,

$$\frac{d}{dx}\psi = \frac{\delta}{2\sqrt{i}\, r(x)} \frac{d}{dx}\theta \quad .$$

The amplitude function $a_o(x)$ in (14) is found from the $O(\omega)$ problem, and the result is, approximately,

$$a_o(x) = \frac{c_o}{\sqrt{A\theta_x}} \quad ,$$

where c_o is an arbitrary constant. Therefore, the general solution of (12a) is

$$p_L(x) \sim \frac{1}{\sqrt{A\theta_x}} \left(c_o e^{i\omega\theta + i\sqrt{\omega}\psi} + c_1 e^{-i\omega\theta - i\sqrt{\omega}\psi}\right) \quad ,$$

where

$$\theta(x) = \int_0^x \sqrt{\frac{K(\xi)}{A(\xi)}}\, d\xi \quad ,$$

and

$$\psi(x) = \frac{\delta}{2\sqrt{i}} \int_0^x \frac{1}{r(x)} \frac{d}{dx}\theta \quad .$$

Applying the two boundary conditions (12b,c) it follows that

$$p_o(x,t) \sim \tfrac{1}{2}\omega\left(\frac{A_oK_o}{AK}\right)^{\frac{1}{4}} \sec(\omega\theta_1+\omega^{\frac{1}{2}}\psi_1) \cdot \frac{A_w}{A_o} \cdot \{e^{i\omega[t-\phi(x)]}-e^{i\omega[t+\phi(x)]}\} \quad , \tag{15}$$

where

$$\phi(x) = \int_x^1 \left[1 + \frac{\delta}{2r\sqrt{i\omega}}\right]\sqrt{\frac{K}{A}} \quad ,$$

$\theta_1 = \theta(1)$, and $\psi_1 = \psi(1)$.

It is clear from (15) that the solution consists of the superposition of left and right traveling waves. Also, since $\phi(x)$ is complex valued, they are attenuated. One can also see that, if the frequency is not too large, the resonating frequencies of the inviscid problem will influence the response. In our case the inviscid modes occur when

$$\cos \omega\theta_1 = 0 \quad ,$$

or equivalently, when

$$\omega_n \int_0^1 \sqrt{\frac{K(x)}{A(x)}}\, dx = \frac{\pi}{2}(2n-1),\ n \in \mathbb{Z}^+ .$$

The effects of these modes on the response will not be pursued here, but the interested reader can find an extended discussion in Holmes (1980b) and Chadwick, et al (1980).

In comparing the attenuation of the two waves in (15) one finds that, for higher frequencies, the right traveling wave dominates. In this case,

$$p_o(x,t) \sim -\omega \frac{A_w}{A_o}\left(\frac{A_oK_o}{AK}\right)^{\frac{1}{4}} e^{i\omega(t-\theta)-i\sqrt{\omega}\psi} \quad . \tag{16a}$$

To obtain this result it is required that

$$e^{2\omega \mathrm{Re}(\phi)} << 1 \quad ,$$

or equivalently,

$$\ln(10) < \delta k_o \sqrt{\tfrac{1}{2}\omega} \int_x^1 \frac{PG^{5/2}}{D^{1/2}A^{3/2}} \quad . \tag{16b}$$

In what follows we deal almost exclusively with the above single wave approximation when discussing the longtime problem. Therefore, it is best to first discuss its applicability, which is indicated by the above inequality. For any location $0 < x < 1$, (16b) shows that (16a) holds for sufficiently high frequencies. However, there is always a region at the apical end where (16b) fails to hold. As the frequency increases the size of this region decreases. Also, note the exponential decay of the wave. Consequently, based on von Bekesy's data for the human cochlea, (16a) serves as a reasonable approximation of the actual solution of (12) for frequencies above about 500 Hz.

Analysis of the Solution of Reduced Problem

One of the central components in the "theory of hearing" is the concept of a place principle. That is, for a given frequency there should be a unique maximum in the amplitude of the wave on the basilar membrane. Using (4) and (16) the centerline displacement of the basilar membrane is, aside from a constant multiple,

$$\eta(x,t) \sim \frac{\omega G^4}{(AK)^{\frac{1}{4}}} e^{i\omega(t-\theta)-i\sqrt{\omega}\psi} . \tag{17}$$

Hence,

$$|\eta| \sim \frac{\omega G^4}{(AK)^{\frac{1}{4}}} e^{-\alpha_0\sqrt{\omega}\, g(x)} , \tag{18}$$

where

$$g(x) = \int_0^x \frac{1}{r(\xi)} \sqrt{\frac{K(\xi)}{A(\xi)}}\, d\xi ,$$

and

$$\alpha_0 = \tfrac{1}{4}\sqrt{2}\delta .$$

Differentiating (18) with respect to x and then setting the result equal to zero one finds that the maximum amplitude occurs when

$$11AG_x - GA_x = k_0\delta\sqrt{2\omega}\, \frac{PG^{7/2}}{A^{\frac{1}{2}}} . \tag{19}$$

Although it is difficult to say much about the solution(s) of (19), note that there is no solution if $G_x < 0$ and $A_x > 0$. However, since the width increases and the cross section decreases in the cochlea this possibility does not arise.

As an example, suppose that $A = A_0$, $P = P_0$, and $G = ax + b$. In this case, the solution x_M of (19) satisfies

$$\sqrt{\omega}\, G_M^{7/2} = \frac{11}{\sqrt{2}} \cdot \frac{aA_o^{3/2}}{\delta k_o P_o} \quad , \tag{20}$$

where

$$G_M = ax_M + b \quad .$$

The values for x_M that are obtained from (20) agree with those found empirically by von Bekesy (Holmes, 1980b).

It is interesting to compare the propagation time $t_p(x)$ of the signal, as found in the transient and longtime problems. From (11)

$$t_p = k_o \int_o^x \sqrt{\frac{G^5}{A}} \quad , \tag{21}$$

and from (17)

$$t_p = k_o \int_o^x (1 + \frac{\delta}{2r\sqrt{2\omega}}) \sqrt{\frac{G^5}{A}} \quad . \tag{22}$$

Based on the order of the expansions used, (21) and (22) are in agreement. They show that the propagation time, and velocity of the wave, are, essentially, independent of frequency. Also, the widening of the basilar membrane, as well as the decreasing of the cross sectional area, act to speed the wave up. Finally, note the sensitivity of the speed on the width of the basilar membrane, as compared to either the cross sectional area or the thickness of the basilar membrane.

As a final topic, we consider the tuning curves one obtains from (18). In contrast to the place principle, these curves are obtained by fixing x and then studying the response as a function of frequency. To determine what frequency produces the maximum response at a fixed spatial location, we can differentiate (18) with respect to ω and then set the result equal to zero. The result is that the frequency ω_M is given as

$$\omega_M = 2\left[\frac{4}{\delta g(x)}\right]^2 \quad . \tag{23}$$

So, ω_M is a monotonically decreasing function of x, which agrees with experiment. Also, it is an increasing function of the width of the basilar membrane, and a decreasing function of the stiffness $D(x)$.

It is interesting to compare (19) and (23). Clearly, they result in different solutions and to see just how different consider the case of when $A = A_o$, $P = P_o$, and $G = ax + b$. From (20) we have that

$$G_M^{7/2}\sqrt{\omega_A} = \frac{11}{\sqrt{2}}\,\frac{aA_o^{3/2}}{\delta k_o P_o} ,$$

and from (23)

$$(G_T^{7/2} - G_o^{7/2})\sqrt{\omega_M} = 14\sqrt{2}\,\frac{aA_o^{3/2}}{\delta k_o P_o} ,$$

where $G_o = G(0)$. Also, ω_A is the frequency used in finding x_M, whereas $G_T = ax_T + b$ represents the half width of the basilar membrane at the spatial location x_T used to obtain ω_M.

Since $G_o^{7/2} << G^{7/2}$, except in the immediate vicinity of the basal end, we have that

$$G_M = (\frac{11}{28})^{2/7} (\frac{\omega_M}{\omega_A})^{1/7} G_T .$$

Thus, taking $G_T = G_M$ it follows that

$$\omega_M = (\frac{28}{11})^2 \omega_A . \qquad (24)$$

To interpret this result, suppose that at a given frequency ω_A the maximum in the amplitude occurs at x_M. The amplitude at x_M in this case is not the maximum amplitude that occurs there. In fact, if one stays at x_M and increases the frequency above ω_A the amplitude continues to increase. It increases until

$$\omega = (\frac{28}{11})^2 \omega_A \sim 6.5\omega_A ,$$

after which it decreases. Clearly, there is a significant difference between the two frequencies. Related to this difference is the question on how the cochlea discriminates the frequency signal. That is, whether it is a raw amplitude analyzer, or whether it is based on the relative, or contrasting, amplitude over the entire membrane. The above analysis would seem to indicate the possibility for both mechanisms.

Conclusions

Now that the low frequency theory has been outlined, a few comments are in order about its overall applicability. It seems, based on von Bekesy's data, that the analysis applies essentially to the lower 25% of the hearing spectrum in humans. Roughly, this means frequencies up to about 1500 or 2000 Hz. In this range the theory results in relatively good agreement for such things as the place principle and the propagation time, although, as mentioned earlier, t_p is independent of frequency in this theory. One aspect it fails to reproduce is the sharpness of the tuning curves that are obtained experimentally (see the article by Khanna and Leonard). There are a number of reasons for this. Among others, it is due to the unidirectional fluid motion, as well as the omission of the inertial effects of the basilar membrane. However, even with these restrictions, or because of them, the low frequency theory has a major advantage over other, more elaborate, theories. This advantage is its simplicity. Because of this it is possible to obtain a simple closed form solution for the wave motion, which illustrates, in a very general way, the dependence on the geometrical components of the cochlea.

Acknowledgement: This work was supported, in part, by the National Science Foundation, Grant No. MCS8102129.

References

von Bekesy, G. (1960). Experiments in Hearing, (McGraw-Hill, New York).

Chadwick, R. S., and Cole, J. D. (1979). "Modes and Waves in the Cochlea," Mech. Res. Comm. 6, 177-184.

Chadwick, R. S., Fourney, M. E., and Neiswander, P. (1980). "Modes and Waves in a Cochlear Model," Hearing Res. 2, 475-483.

Helmholtz, H. (1885). On the Sensations of Tone, (Dover, New York).

Holmes, M. H. (1980a). "Low Frequency Asymptotics for a Hydroelastic Model of the Cochlea," SIAM J. Appl. Math 38, 445-456.

Holmes, M. H. (1980b). "An Analysis of a Low-Frequency Model of the Cochlea," J. Acoust. Soc. Am. 68, 482-488.

Holmes, M. H. (1981). "Study of the Transient Motion in the Cochlea," J. Acoust. Soc. Am. 69, 751-759.

Holmes, M. H. (1982). "A Mathematical Model of the Dynamics of the Inner Ear," accepted for publication in J. Fluid Mech.

Peterson, L. and Bogert, B. (1950). "A Dynamical Theory of the Cochlea," J. Acoust. Soc. Am. 22, 369-381.

Zwislocki, J. (1965). "Analysis of Some Auditory Characteristics," Handbook of Math Psychology 3, 1-97.

BASILAR MEMBRANE RESPONSE MEASURED IN DAMAGED COCHLEAS OF CATS

S. M. Khanna and D. G. B. Leonard
Department of Otolaryngology
Columbia University
630 West 168th Street
New York, New York 10032

INTRODUCTION

A variety of techniques have been used in the past to measure the vibrations of the basilar membrane in different animals (Bekesy, 1947, stroboscopic microscope; Johnstone and Boyle, 1967, Mössbauer effect; Wilson and Johnstone, 1972, 1975, capacitive probe; Kohllöffel, 1972, fuzziness detection). The present method utilizes an interferometric technique with high sensitivity. Basilar membrane vibrations were measured in cats utilizing this technique.

METHODS

The Interferometric Technique and Measurement. The measurement of basilar membrane vibrations was carried out using the interferometric comparison method as described by Khanna (1981). The experimental method is outlined briefly below.

The cochlea was surgically opened and a mirror placed on the basilar membrane. Light from a Helium Neon laser was focussed onto the basilar membrane mirror. A reference mirror was positioned in the path of the laser beam and then cemented to the skull of the cat. The reflections from the two mirrors were adjusted so that they passed through an aperture onto a photomultiplier, which detected the intensity of the incident light. Vibration of one or both mirrors produced an ac output from the photomultiplier.

Vibration of the basilar membrane mirror was produced by sound applied to the tympanic membrane of the cat using a driver designed by Sokolich (1977). Sound pressure levels at the tympanic membrane were measured using a probe microphone also designed by Sokolich (1977). This combination allowed us to apply and measure sound pressures very precisely up to frequencies of 40 kHz. The harmonic margin of the sound system was greater than 60 dB.

A comparison method was utilized to determine the absolute vibration amplitude of the basilar membrane mirror. The reference mirror was vibrated at a known frequency and amplitude. The magnitudes of displacement of the basilar membrane mirror and reference mirror were related to the photomultiplier ac output at the sound frequency and reference frequency, respectively. Therefore, the absolute vibration amplitude of the basilar membrane mirror was calculated from the ac outputs and the known vibration of the reference mirror. This interferometric method is extremely sensitive. The minimum vibration amplitude that can be measured at present is 3×10^{-11} cm with a 20 dB signal-to-noise ratio. The maximum vibration amplitude that can be measured is 10^{-5} cm. This results in a dynamic range of over 100 dB. The frequency response of the interferometric measurement system is within 3 dB from 100 Hz to 30 kHz. The measurements are very repeatable with time (within a fraction of a dB) and the method allows for phase as well as transient response measurements.

Surgical Technique. Two criteria were used in selecting the position for opening scala tympani for access to the basilar membrane. First, the frequency of tuning of the basilar membrane should be as low as possible. Therefore, the hole should be as far apical as possible. Second, the plane of the exposed basilar membrane should be perpendicular to the axis of the hole. This was necessary to insure that the reflection from the basilar membrane mirror would return through the scala tympani opening.

The method of locating and drilling such a hole is based upon a technique developed by Dr. J. Tonndorf. The hole was made in the bone just apical to the round window. This exposed an area of basilar membrane approximately 4 to 5 mm from the basal end of the cochlea. Dr. Tonndorf's method had only been used for cadaver cats. The following is a brief outline of his surgical method. A step was drilled into the posterior part of the petrous bone, apical to the round window, without perforating the vestibule, cochlear aqueduct, scala tympani or the brain cavity. This step was then widened posteriorly using dental cement to accommodate a 2.5 mm diameter stainless steel tube. The tube was cemented to the step such that the axis of the tube was perpendicular to and overlying the basilar membrane. Working through the tube, the cochlear bone was completely drilled away, leaving the endothelial lining of scala tympani intact. The bone chips and dust were removed as much as possible and the endothelium was cut to make the cochlear opening. The final result is diagrammed in Figure 1.

This surgery usually failed due to one of the following problems.

1) While making the step into the bone, a perforation was made into the vestibule, the cochlear aqueduct, scala tympani or the brain cavity.

2) Due to moisture from the surrounding tissues, the cement did not adhere to the cochlear bone.

3) The tube position or its axis was such that the basilar membrane

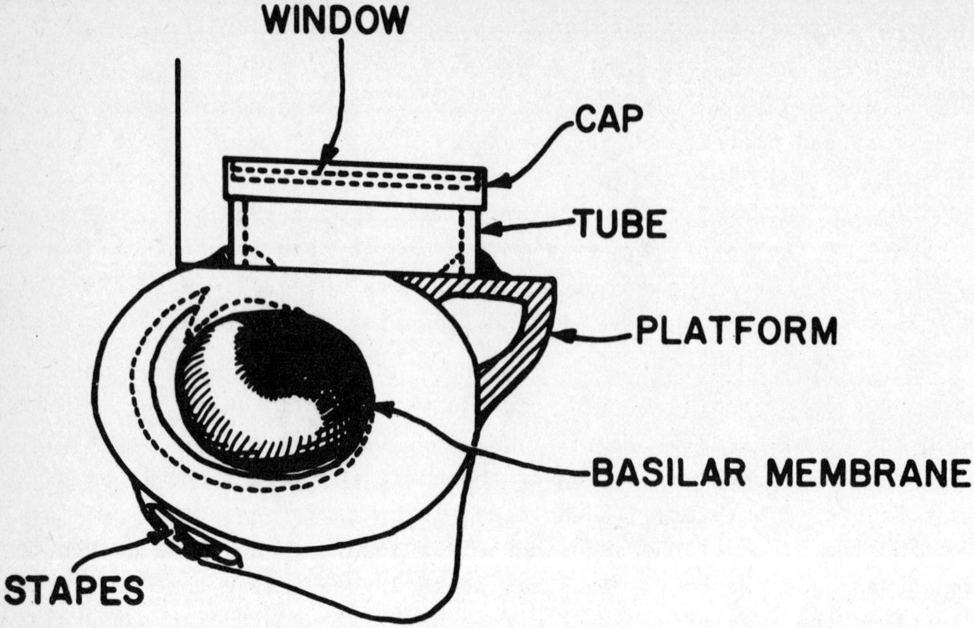

Fig. 1. Surgical opening of scala tympani for access to the basilar membrane. A step was drilled in the petrous bone. The area of the step was increased to accommodate a stainless steel tube which was cemented onto the step. Working through the tube, scala tympani was opened. After placing the mirror on the basilar membrane, the end of the tube was closed with a fluid tight glass window.

could not be visualized properly.

4) While cutting the endothelium, bone dust or chips and other debris would fall onto the basilar membrane.

This surgical method was adapted by us for experiments on live cats. Improvements and alterations were made in the technique until all the above problems were solved. However, working with living cats presented new difficulties. Initially, the most severe problem was bleeding from vessels in the endothelium. Even a small amount of blood entering the cochlea caused the formation of a thin fiber network in scala tympani. The presence of these fibers prevented the mirrors used for measurement from reaching the basilar membrane. Cauterization of the blood vessels appeared to be the only satisfactory, practical method for avoiding or stopping bleeding. However, commercial cautery machines produced excessive heat and consequently damaged the cochlea.

A thermal microcautery was designed and built. In addition, a two-step cauterization procedure was used. First, the blood vessels were cauterized through a very thin, intact layer of bone. Then, after the endothelium was exposed, the blood vessels were recauterized as needed. Once bleeding could be controlled, we were able to place mirrors on the basilar membrane.

Mirror Placement. The basilar membrane mirrors used in our experiments were chemically grown gold crystals. They were hexagonally shaped plates approximately 10^{-6} cm thick and between 75 and 100 μ across; they weighed less than 10^{-8} gm.

The mirrors were handled with a polyvinyl chloride microtip attached to a micromanipulator system. A mirror was picked up by bringing the fluid filled microtip near the mirror and applying a negative fluid pressure. The microtip and mirror were then transferred to the cochlea through the hole into scala tympani. The mirror was then released from the microtip and floated down to the basilar membrane. Scala tympani was not drained during this procedure.

After placing the mirror, the cochlear opening was closed by a fluid tight cap with an optical window. The direction of the reflection from the basilar membrane mirror was adjusted by rotating the head of the animal until the beam passed through a small aperture and onto the photomultiplier. Then the reference mirror was positioned and cemented to the cat's skull. It was at this stage that we encountered a second serious difficulty.

Mirror Reflection Movements. The direction of the mirror reflections was expected to be stable and the success of the interferometry depended on the two beams staying on the photomultiplier. Unfortunately, large, slow, angular movements of the basilar membrane mirror (on the order of 5 to 10 degrees) caused the reflected beam to move off of the photomultiplier. Once this happened, the beam had to be brought back by rotating the animal's head. This rotation moved the reference beam off the photomultiplier, so it too had to be realigned. Measurements could be taken only as long as the two beams stayed on the photomultiplier. This duration was usually two to thirty minutes at the beginning of an experiment, decreasing with time. The experiments were terminated when the beams did not stay in position long enough to carry out accurate measurements. When the mirror movements were present at the beginning of an experiment, usually no measurements could be made. In only about 10% of the experiments were stable reflections obtained at the start of an experiment. Even so, observation time was limited because the mirrors lost their light reflectivity and scattered the light within a few hours of being placed in the cochlea.

There were several independent reasons for the mirror movements. We found that even a small amount of blood would produce a fiber network in scala tympani. In addition, blood entering the cochlea collected on the basilar membrane, creating fibers directly on the basilar membrane. A mirror resting on one of these fine fibers would move as the perilymph flowed through scala tympani. Frequently, the presence of such thin fibers was discovered while dropping the mirrors. As the microtip was withdrawn from scala tympani after dropping a mirror, a fiber would be seen attached to the tip. All of the mirrors already on the basilar membrane would be pulled out with the fiber as the microtip continued to be withdrawn.

During several experiments, a very fine, almost transparent membrane was found either floating in the perilymph or lying on the basilar membrane. It was thought to be the innermost layer of the endothelial membrane which separated from the thicker layer and collapsed onto the basilar membrane. Often, the mirrors were dropped onto this membrane. In all such cases, the mirror reflections drifted. A fiber-optic illuminator was built to illuminate the basilar membrane from below and improve our ability to detect such membranes.

A third cause of mirror movements was bacteria. The mirrors were stored in water. It seems that bacteria were growing on them. The growth of the bacterial colonies around the mirrors on the basilar membrane caused mirror movements. This source was discovered in a later series of experiments (Khanna and Leonard, 1981 B).

Damage Assessment. Once it became possible to make interferometric measurements in living cats, we felt that it was essential to evaluate the condition of the cochlea after surgical preparation. A two-fold approach was taken. First, the cochleas were examined histologically for damage. In a series of experiments, cochleas were processed as epon-embedded surface preparations (Liberman and Beil, 1979) after various stages of the surgery, and then examined for damage. For each cat, the cochlea with no surgery was used as a control. Damage to the operated cochlea was found in each case. A manuscript describing these results is in preparation (Leonard and Khanna, 1981).

Secondly, since interferometric measurements were being made close to the round window, it was possible to assess cochlear damage by measuring round window cochlear microphonics (CM). Pure tones from 125 Hz to 30 kHz were used as stimuli for the CM measurements. CM measurements were made after each surgical step. The total CM loss often exceeded 60 dB. By both damage assessment techniques, it was found that drilling and cauterizing were damaging the cochlea. (For details of factors which produce cochlear damage see Khanna and Leonard, 1981 A.)

Two factors could produce damage during drilling: (1) noise due to the drill and/or bone vibrations; and (2) heat due to friction. Noise levels were measured for various types of dental drills, and all were found to be excessive for our application. Therefore, a drill was designed and built which would have minimal vibration and noise levels at least 30 dB below the drill used previously. In addition, air at 0° C was blown over the drill bit to reduce the heat. Simultaneously, a new microcautery was designed which reduced the total heat output by reducing the duration of each heat impulse to less than 100 msec.

In spite of these and many other improvements, the CM loss produced by this experimental procedure could not be reduced below 10 to 15 dB in the frequency region above 20 kHz. In addition, such reduced CM loss was achieved in only a few experiments. It was clear that the surgical

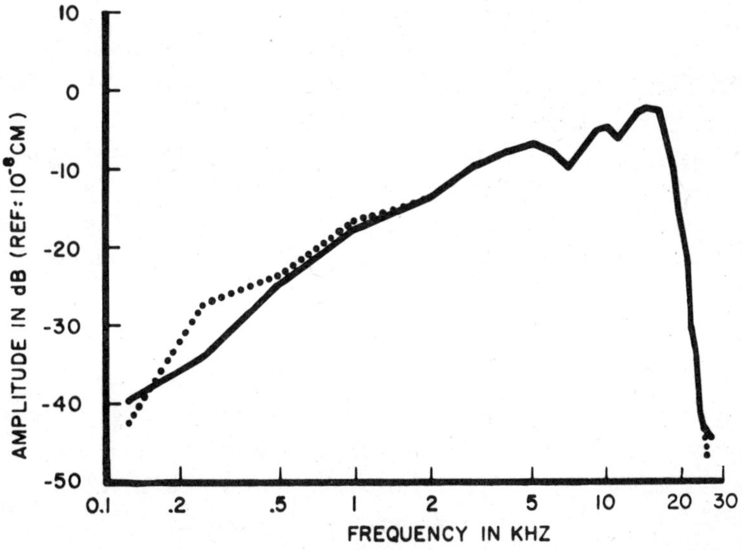

Fig. 2. Relative amplitude of basilar membrane vibrations as a function of frequency for 60 dB SPL at the tympanic membrane. Close agreement between two sets of observations taken an hour apart shows the excellent repeatability of the method (cat, 5-29-79).

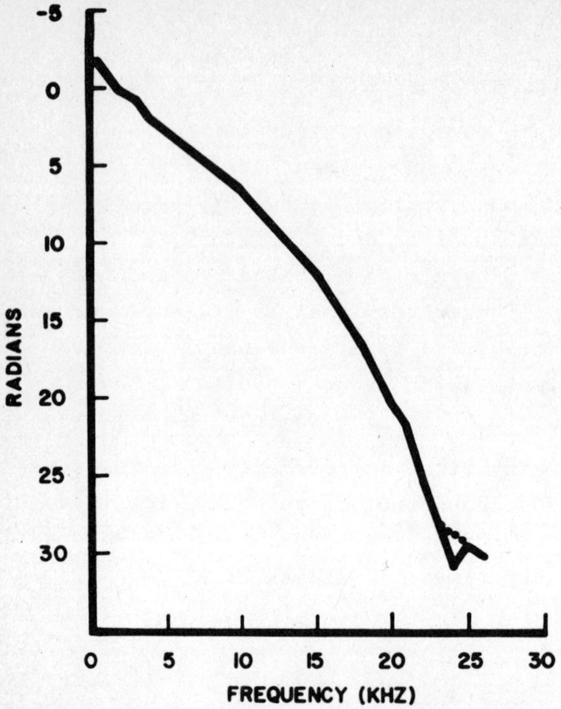

Fig. 3. Phase of the basilar membrane vibrations with reference to the phase of the sound pressure at the tympanic membrane. Two sets of data shown as a function of frequency are within 0.5 radians, showing that phase can be measured with excellent repeatability (cat, 5-29-79).

technique produced extensive damage to the cochlea. Therefore, this surgical method was abandoned. A new method was developed in which the basilar membrane was accessed by cutting the round window membrane (Khanna and Leonard, 1981 B). Further investigations showed that trauma to the cochlea can be produced by acoustical, mechanical, thermal, chemical, light or electrical stimuli (Khanna and Leonard, 1981 A).

RESULTS

The excellent repeatability of the method is demonstrated by plots of two successive sets of amplitude measurements (Fig. 2) and phase measurements (Fig. 3). The relative amplitude of basilar membrane vibration as measured for constant sound pressure level at the tympanic membrane is plotted versus frequency in Figure 2. (These data were obtained before the calibration technique used for obtaining absolute magnitude was fully developed (Khanna, 1981).) The two sets of measurements are within a decibel of each other between 500 Hz and 26 kHz. Two sets of phase measurements from the same experiment are shown in Figure 3. The phase values are within 0.5 radians of each other for each frequency measured. Basilar membrane vibration amplitude as a function

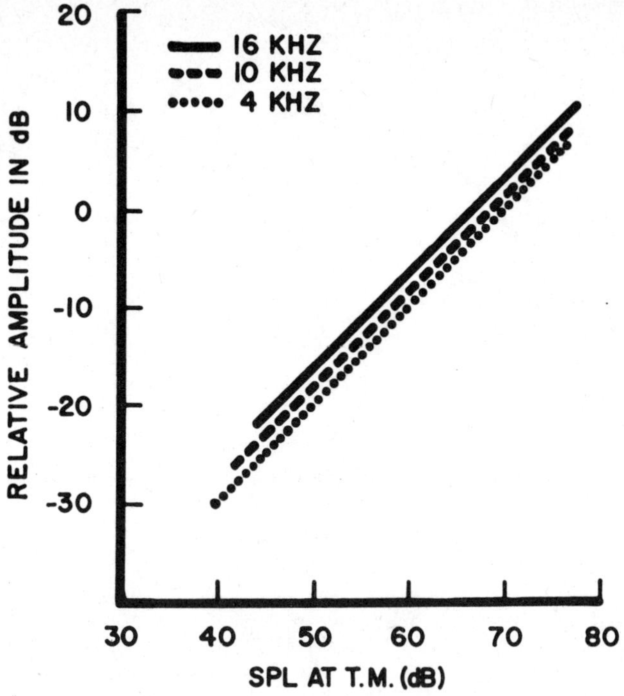

Fig. 4. Relative amplitude of basilar membrane vibration as a function of sound pressure level measured at 4, 10 and 16 kHz. Basilar membrane vibrations increase linearly with sound pressure. (cat, 5-29-79)

of sound pressure level is shown in Figure 4 for three frequencies (4 kHz, 10 kHz and 16 kHz). The amplitude increases linearly with sound pressure level at each of the three frequencies. The vibration amplitude can be measured at sound pressure levels as low as 40 dB. No data about the condition of this cochlea are available. The CM measurements and histological assessments were carried out only in later experiments. Since all our effort to reduce the damage to the cochlea began later, we can only conclude that this cochlea was severely damaged.

Basilar membrane vibration measurements were obtained after the reduction of the surgical trauma by the improvements described above. The comparison method of measurement was also fully implemented to obtain the absolute magnitude of vibration. The absolute amplitude of basilar membrane vibration as a function of frequency is shown in Figure 5. Two sets of data are shown. The first is for a sound pressure level of 70 dB at the tympanic membrane and the second is at 80 dB. The measurements of

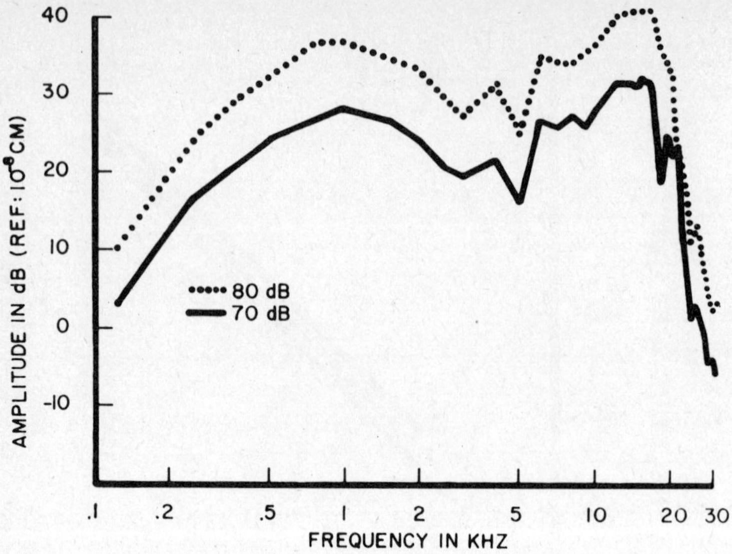

Fig. 5. Absolute amplitude of basilar membrane vibrations as a function of frequency and sound pressure level. Peak amplitude is shown for 70 and 80 dB r.m.s. SPL at the tympanic membrane. An increase of 10 dB SPL produced an increase of only 8 dB in vibration amplitude at most frequencies. (cat, 2-15-80)

displacement at this time could not be obtained at lower sound pressure levels. The interferometric technique was later improved so that measurements could be obtained at sound pressure levels as low as 30 dB. With the exception of the frequency region above 17 kHz, the shape of the two curves is nearly identical. The 80 dB SPL curve superimposes on the 70 dB SPL curve if it is shifted down by 8 dB. An increase of sound pressure by 10 dB thus increases the vibration amplitude by only 8 dB. The basilar membrane response shows a broad maximum between 125 Hz and 3 kHz. The initial slope at 125 Hz is approximately 13.5 dB/octave, and the maximum is located at 1.0 kHz. A second broad maximum is located in the frequency region of 5 to 30 kHz with a peak between 12 and 13 kHz. No sharp peak is seen in this basilar membrane tuning curve. We were to learn later that the response in the high frequency peak region is extremely sensitive to trauma, as is the response in the high frequency cutoff region between 20 and 30 kHz (Khanna and Leonard, 1981 C). In addition, the high sound pressure levels used for this interferometric technique are partly responsible for the flat frequency response due to the non-linearities of the basilar membrane vibrations. However, these amplitudes are very similar in shape to those obtained

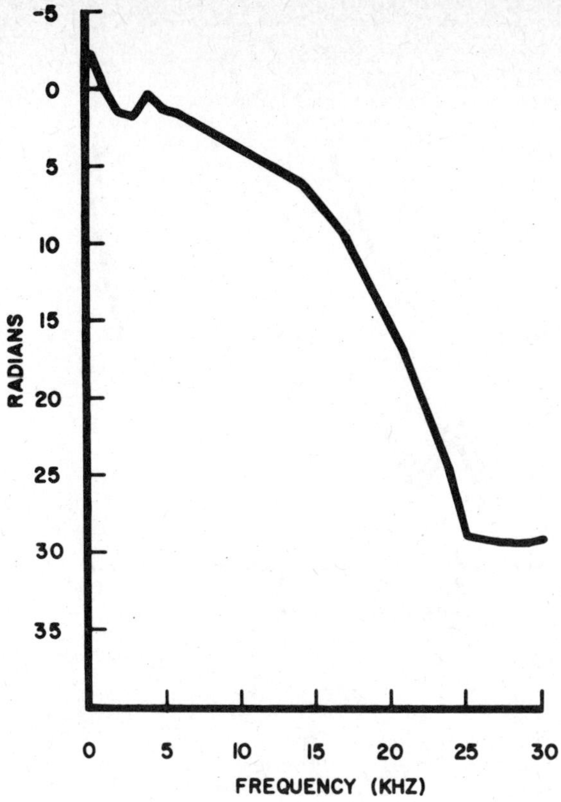

Fig. 6. Basilar membrane phase measured with respect to the phase of sound pressure (80 dB SPL) at the tympanic membrane. The initial slope between 4 and 15 kHz is -0.6 radians/kHz, it increases to -1.88 radians/kHz between 18 and 21 kHz, and to -2.5 radians/kHz between 21 and 25 kHz. A phase plateau of 29.5 radians is observed at frequencies above 25 kHz. (cat, 2-15-80)

by Evans and Wilson (1975) in cat and to those obtained by Wilson and Johnstone (1975) in guinea pig.

The basilar membrane phase, measured with reference to the phase of the sound pressure at the tympanic membrane, is shown in Figure 6. Our phase curves include the contribution of the middle ear. The contribution of the middle ear phase shifts can be obtained from the data of Guinan and Peake (1967) which was measured under the same conditions as our data (open bulla, septum removed). Above 4 kHz, our data show a steadily increasing phase lag with increasing frequency. The slope of the curve between 4 and 14 kHz is -0.6 radians/kHz. Between 17 and 21 kHz the slope is -1.88 radians/kHz, and between 21 and 25 kHz it is -2.5 radians/kHz. Abruptly at 25 kHz, the phase plateaus at 29.5 radians. The phase lag remains constant as the frequency is increased to 30 kHz. Similarly shaped basilar membrane phase curves are observed by Rhode (1978) in squirrel monkey. His phase slopes are steeper (-2.9 and -9.4 radians/kHz), but the value of the phase plateau is nearly the same

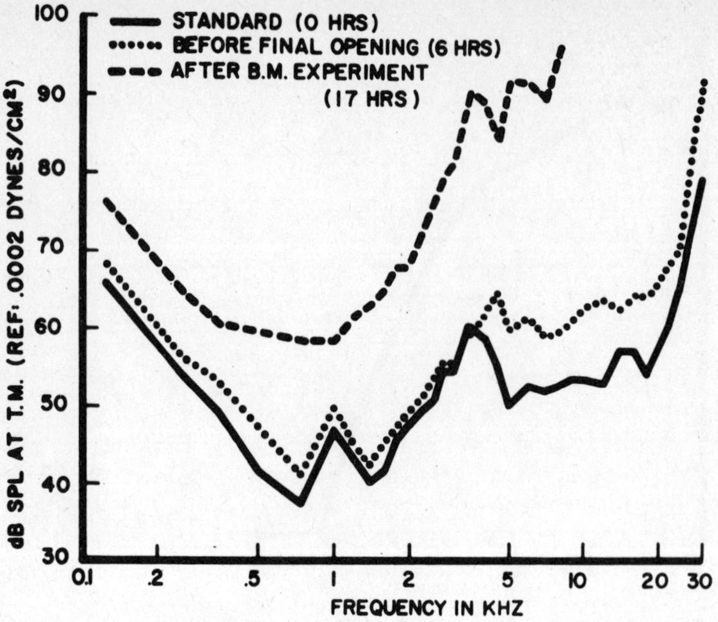

Fig. 7. Assessment of cochlear damage using round window cochlear microphonics (CM). The r.m.s. sound pressure levels required at the tympanic membrane to produce a 10 μv r.m.s. CM response shown as a function of frequency (1) in the normal bulla open condition, (2) after completion of drilling and cauterizing but before the final opening of the cochlea, and (3) after completion of the basilar membrane experiment. These curves show that the surgical procedure produced extensive damage to the cochlea. (cat, 2-15-80)

(28.2 radians).

The condition of this cochlea was assessed by measuring CM. The sound pressure level required to produce a 10 μv CM response was measured (1) in the bulla open condition before drilling and cauterizing, (2) after completion of drilling and cauterizing but before the final opening of the cochlea, and (3) after completion of the basilar membrane vibration measurements. These curves are shown in Figure 7. Comparison of the curves shows that drilling and cauterizing produced extensive damage to the cochlea. Between 125 Hz and 3 kHz, the loss was between 2 and 4 dB. Above 5 kHz, the loss was between 6 and 10 dB. CM measurements taken immediately after the completion of the basilar membrane vibration measurements show even more CM loss (10 dB to 300 Hz; 20 dB between 300 Hz and 2 kHz; 30 dB at 5 kHz). The CM response was not measurable above 8 kHz; therefore, we can only conclude that the loss exceeded 40 dB.

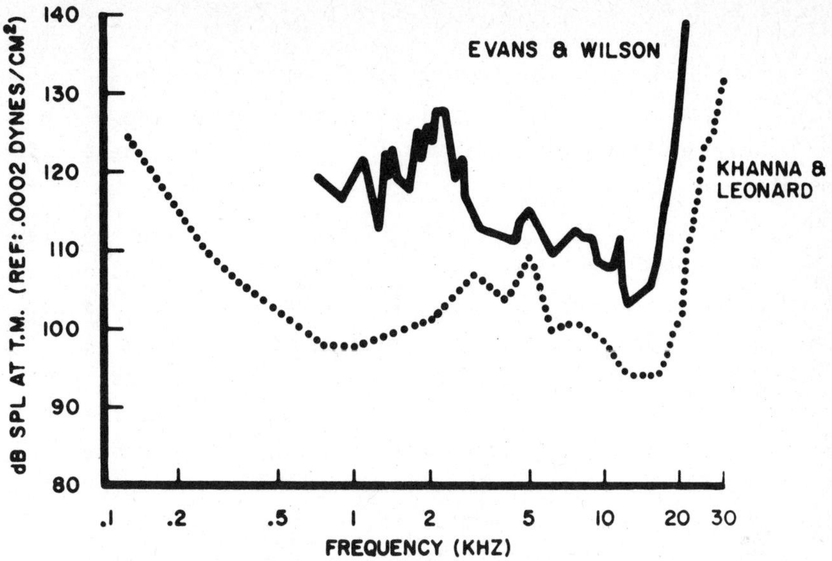

Fig. 8. Comparison of data. The r.m.s. sound pressure required at the tympanic membrane to produce 400 Å r.m.s. basilar membrane displacement in cat. (1) Evans and Wilson (1975), (2) Khanna and Leonard (cat, 2-15-80). The general shapes are quite similar and the absolute magnitudes of vibration agree within approximately 10 dB in the frequency region above 3 kHz.

DISCUSSION

A comparison of our data (from Figure 5) and the data of Evans and Wilson (1975) is shown in Figure 8. Both sets of data were obtained in cats, and from a similar frequency region of the basilar membrane. Our data were from a location about 5 mm from the basal end of the basilar membrane, while Evans and Wilson measured at 6.5 mm from the basal end. Our data has been linearly extrapolated to obtain the sound pressure required for a r.m.s. displacement of 400 Å, so that a direct comparison could be made. The curves are very similar. Above 3 kHz, the absolute magnitudes are within 10 dB, and both have a shallow negative peak between 12 and 17 kHz. The difference in the cutoff frequencies is due to the slightly different locations at which the measurements were made.

Although the basilar membrane frequency responses from the two experiments are similar, the conclusions about the condition of the two cochleas are not. Our experiments show that there was extensive CM loss. Therefore, we conclude that the cochlea must have been damaged. Evans and Wilson have argued that the cochlea in their experiment was not damaged because they measured normal neural tuning curves with characteristic frequencies of tuning at the same frequency as the peak of the basilar membrane tuning curve. The key differences between these two experiments are the methods of evaluating the condition of the cochlea and the interpretations of these results. Therefore, these differences will be discussed in detail.

An alteration in the basilar membrane mechanical properties due to trauma will produce a change in the hair cell response as well as in the neural response. However, a change in the CM response does not prove that the mechanical response of the basilar membrane has been altered. The CM change may be due to damage of one of the elements involved in the production of CM, while the mechanical response may remain unchanged. Wilson and Johnstone (1975) found that the mechanical properties of the basilar membrane did not vary in any way with the physiological condition as judged by gross neural potential (N_1) thresholds. This would indicate that physiological changes of the cochlea do not produce changes in the basilar membrane mechanics.

However, in our later experiments using a surgical approach through the round window, the CM loss was substantially reduced and changes were observed in the mechanical response of the basilar membrane (Khanna and Leonard, 1981 C). It should be noted that most changes in the mechanical response took place only when the CM loss was less than a few dB in magnitude. Rhode (1978) has also measured basilar membrane responses (in squirrel monkey) which are much sharper than those observed by Evans and Wilson in cat.

Our experiments have shown that trauma reduces the peak amplitude, broadens the basilar membrane frequency response and shifts the peak to a lower frequency region (Khanna and Leonard, 1981 C). Due to both broadening and shifting, it would be very difficult to identify the nerve fibers which originated from the region of damage. Moreover, the sensitivity of the neural tuning curve would be so drastically reduced that it would be very difficult to find these fibers. Therefore, we conclude that Evans and Wilson probably recorded from nerve fibers originating in a more apical region of the basilar membrane than where the basilar membrane responses were measured.

It is not commonly recognized that the cochlea is extremely susceptible to trauma from a variety of stimuli. Gross damage may be detected by monitoring changes of the CM or N_1 response. However, small, localized damage is often not detected using these techniques.

When measuring the basilar membrane frequency response, many of the earlier investigators did not monitor the condition of the cochlea (Bekesy, 1947; Johnstone and Boyle, 1967; Rhode, 1971). The basilar membrane was classically considered as a mechanical system. There was no reason to believe that the response of a purely mechanical system would be affected by the physiological condition of the cochlea. Later, when the physiological condition of the cochlea was monitored, signs of gross damage were seen (Wilson and Johnstone, 1975; Lepage and Johnstone, 1980). However, the mechanical response did not seem to be related to the observed changes in the N_1 response. This strengthened the concept that the mechanical response of the basilar membrane did not depend on the physiological condition of the cochlea.

Changes in the basilar membrane response were not observed because the cochlea was already severely damaged at the time of the initial measurements. Since trauma affects the response mainly in the peak region (Khanna and Leonard, 1981 C), once the peak is flattened, additional damage does not produce appreciable changes in the observed response. Such a damaged response was measured by us in the present series of experiments.

It is our suggestion that the mechanical response of an undamaged cochlea has not yet been measured.

REFERENCES

Bekesy, G. von (1947) The vibration of the cochlear partition in anatomical preparations and in models of the inner ear. J. Acoust. Soc. Am. 21 233-245.

Evans, E. F. & Wilson, J. P. (1975) Cochlear tuning properties: concurrent basilar membrane and single nerve fiber measurements. Science 190 1218-1221.

Guinan, J.J., Jr. & Peake, W. T. (1967) Middle-ear characteristics of anesthetized cats. J. Acoust. Soc. Am. 41 1237-1261.

Johnstone, B. M. & Boyle, A. J. (1967) Basilar membrane vibrations examined with the Mössbauer technique. Science 158 390-391.

Khanna, S. M. (1981) A comparison technique of interferometric measurements. In preparation.

Khanna, S. M. & Leonard, D. G. B. (1981 A) Cochlear damage incurred during preparation for and measurement of basilar membrane vibrations. In preparation.

Khanna, S. M. & Leonard, D. G. B. (1981 B) Basilar membrane vibrations measured in cat using a round window approach. In preparation.

Khanna, S. M. & Leonard, D. G. B. (1981 C) Basilar membrane tuning in the cat cochlea. In preparation.

Kohllöffel, L. U. E. (1972) A study of basilar membrane vibrations II. The vibratory amplitude and phase pattern along the basilar membrane (post-mortem). Acustica 27 68-81.

Leonard, D. G. B. & Khanna, S. M. (1981) Histological evaluation of damage in cat cochleas used for measurement of basilar membrane mechanics. In preparation.

Lepage, E. L. & Johnstone, B. M. (1980) Nonlinear mechanical behavior of the basilar membrane in the basal turn of the guinea pig cochlea. Hear. Res. 2 183-189.

Liberman, M. C. & Beil, D. G. (1979) Hair cell condition and auditory nerve response in normal and noise-damaged cochleas. Acta Otolaryng. 88 161-176.

Rhode, W. S. (1971) Observations of the vibration of the basilar membrane in squirrel monkey using the Mössbauer technique. J. Acoust. Soc. Am. 49 1218-1231.

Rhode, W. S. (1978) Some observations on cochlear mechanics. J. Acoust. Soc. Am. 64 158-176.

Sokolich, W. G. (1977) Improved acoustic system for auditory research. J. Acoust. Soc. Am. 62, Suppl. 1 S21 (abstract).

Wilson, J. P. & Johnstone, J. R. (1972) Capacitive probe measures of basilar membrane vibration. Symposium on Hearing Theory, IPO Eindhoven, Holland 172-181.

Wilson, J. P. & Johnstone, J. R. (1975) Basilar membrane and middle ear vibrations in guinea pig measured by capacitive probe. J. Acoust. Soc. Am. 57 705-723.

A Mathematical Model of the Semicircular Canals

by
William C. Van Buskirk
Department of Biomedical Engineering
Tulane University
New Orleans, Louisiana 70118

1. Introduction

The semicircular canals are the primary transducer for the sensing of angular motion. As such, they are part of the organs of equilibrium. The importance of these organs for the successful functioning of the human body is obvious. The semicircular canals have, therefore, attracted the attention of physiologists, sensory psychologists and physicians over the years. From the very beginning, physical scientists and mathematicians have been consulted to provide an explanation of the mechanics of this fascinating organ.

The advent of aerospace flight with its new demands on the human organism has accelerated the pace of vestibular research. It has become apparent that, while the semicircular canals are an engineering system of some elegance, they are capable of producing disorienting sensations when subjected to non-physiological motion. This knowledge has helped aerospace planners to avoid situations which might prove discomforting or disabling to the pilot or astronaut.

This alone is an adequate reason for wishing to develop a mathematical model of the semicircular canals, but a further incentive comes from the fact that a full understanding of the mechanics of healthy semicircular canals may contribute to the diagnosis and treatment of canals in a diseased state.

2. Anatomy and physiology

The semicircular canals are located, along with the organ of hearing, in the inner ear. There are three sets of canals on each side of the head (see Figure 1). They are oriented in almost mutually orthogonal planes so that rotation about any axis may be properly sensed. As shown, each canal consists of two parts: an outer canal, which is a channel carved in bone, and an inner, membranous canal. The inner canal is filled with a fluid called *endolymph*. The space between the membranous and bony canal is filled with *perilymph*, a fluid different in composition from endolymph.

One end of each semicircular *duct* is enlarged to form its *ampulla*. The ampulla nearly fills the cross-section of the bony canal and terminates on the *utricle*. The ampulla contains the *cupula*, a gelatinous dividing partition with the same density as endolymph. The cupula fills the entire cross-section of the ampulla, thus interrupting the otherwise continuous fluid path through the duct, utricle and ampulla.

The cupula is the system transducer. It is connected to nervous tissue at its base. Mechanical deflection of the cupula is converted into electrical impulses which transmit the state of angular motion along the vestibular nerve to the central nervous system.

Qualitatively, the manner in which the semicircular canals work is as follows. An angular acceleration of the head causes the bony canals and the membranous structure attached to them to accelerate in a similar manner. The inertia of the endolymph, however, causes it to lag behind the motion of the head. Thus there is a flow of endolymph relative to the duct walls. This flow deflects the cupula, initiating the electrical impulses to the brain.

3. Formulation of the problem

In this **section**, we develop a mathematical model for fluid flow in a single semicircular canal. The membranous semicircular canal duct is approximated by a section of a rigid torus filled with an incompressible Newtonian fluid. For the

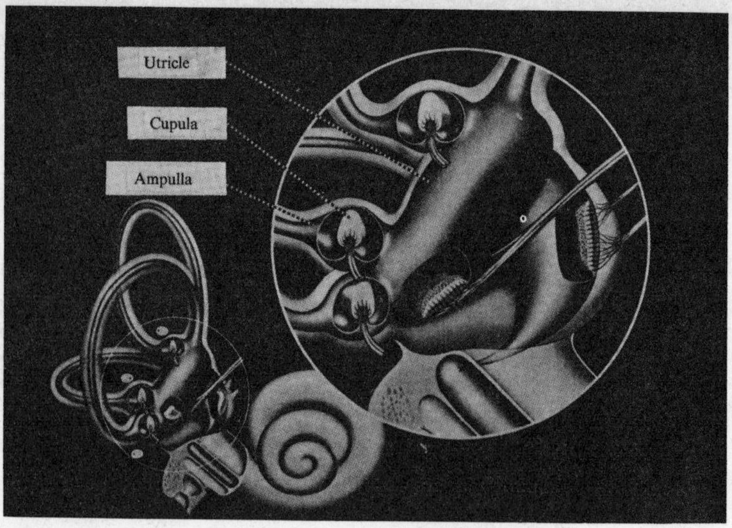

FIGURE 1. The Semicircular Canal.

purpose of this analysis, the perilymph is assumed to have no effect on the deflection of the cupula. The governing equation for the flow of fluid in the duct is the classical Navier-Stokes equation

$$\frac{\partial \underset{\sim}{v}}{\partial t} + \underset{\sim}{v} \cdot \nabla \underset{\sim}{v} = - \frac{1}{\rho}\nabla p + \underset{\sim}{b} + \nu \nabla^2 \underset{\sim}{v} \tag{1}$$

where $\underset{\sim}{v}$ is the velocity of the fluid with respect to an inertial reference frame, ρ is the density, p is the pressure, $\underset{\sim}{b}$ is the body force and ν is the kinematic viscosity.

We are interested in the flow of the fluid with respect to the duct. Therefore we introduce the symbol u, which will represent the velocity of the fluid relative to the duct wall. If $\underset{\sim}{v}_w$ is the velocity of the wall, then

$$\underset{\sim}{v} = \underset{\sim}{u} + \underset{\sim}{v}_w. \tag{2}$$

Now the velocity of a given point on the duct wall is given by

$$\underset{\sim}{v}_w = \underset{\sim}{v}_c + \underset{\sim}{\omega} \times (\underset{\sim}{R} + \underset{\sim}{r}), \tag{3}$$

where $\underset{\sim}{v}_c$ is the velocity of the center of curvature of the duct, $\underset{\sim}{\omega}$ is the angular velocity of the canal, $\underset{\sim}{R}$ is the position vector of the center of the duct with respect to the center of curvature, and $\underset{\sim}{r}$ is the position vector of the point on the duct wall with respect to the center of the duct (see Figure 2). Since $|\underset{\sim}{r}|/|\underset{\sim}{R}| \ll 1$, we can approximate $\underset{\sim}{v}_w$ by

$$\underset{\sim}{v}_w \simeq \underset{\sim}{v}_c + \underset{\sim}{\omega} \times \underset{\sim}{R}. \tag{4}$$

Therefore

$$\underset{\sim}{v} = \underset{\sim}{u} + \underset{\sim}{v}_c + \underset{\sim}{\omega} \times \underset{\sim}{R}. \tag{5}$$

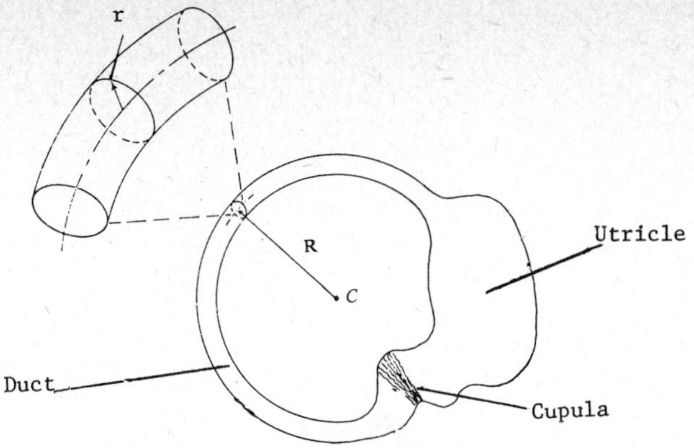

FIGURE 2. A schematic diagram of a single semicircular canal. $\underset{\sim}{R}$ is a vector from C, to the center of the duct and $\underset{\sim}{r}$ is a vector from the tip of $\underset{\sim}{R}$ to the duct wall.

Substituting (5) into (1) we obtain

$$\frac{\partial \underset{\sim}{u}}{\partial t} + \underset{\sim}{u} \cdot \nabla \underset{\sim}{u} + \underset{\sim}{a}_c + \underset{\sim}{\alpha} \times \underset{\sim}{R} + \underset{\sim}{\omega} \times (\underset{\sim}{\omega} \times \underset{\sim}{R}) = -\frac{1}{\rho}\nabla p + \underset{\sim}{b} + \nu \nabla^2 \underset{\sim}{u}, \tag{6}$$

where $\underset{\sim}{a}_c$ is the acceleration of the center of curvature of the duct and $\underset{\sim}{\alpha}$ is the angular acceleration of the canal.

Let the pressure gradient ∇p be split into three parts:

$$\nabla p = \nabla p' + \nabla p'' + \nabla p''', \tag{7}$$

where $\nabla p''$ and $\nabla p'''$ are the pressure gradients associated with $\underset{\sim}{a}_c$ and $\underset{\sim}{b}$ (which for this problem is $\underset{\sim}{g}$, the vector acceleration of gravity); i.e.

$$\underset{\sim}{a}_c = -\frac{1}{\rho}\nabla p'' \tag{8}$$

and

$$\underset{\sim}{b} = \underset{\sim}{g} = \frac{1}{\rho}\nabla p'''. \tag{9}$$

The body force (gravity) and the acceleration $\underset{\sim}{a}_c$ lead to no flow in the canal.

Cancelling the terms indicated by (8) and (9) we obtain

$$\frac{\partial \underset{\sim}{u}}{\partial t} + \underset{\sim}{u} \cdot \nabla \underset{\sim}{u} + \underset{\sim}{\alpha} \times \underset{\sim}{R} + \underset{\sim}{\omega} \times (\underset{\sim}{\omega} \times \underset{\sim}{R}) = -\frac{1}{\rho}\nabla p' + \nu \nabla^2 \underset{\sim}{u}. \tag{10}$$

Since $|\underset{\sim}{r}|/|\underset{\sim}{R}| << 1$, we may express this equation in cylindrical co-ordinates. The axial component is then

$$\frac{\partial u}{\partial t} + R\alpha = -\frac{1}{\rho}\frac{\partial p}{\partial z} + \frac{\nu}{r}\frac{\partial}{\partial r}\left(r\frac{\partial u}{\partial r}\right), \tag{11}$$

where r is the radial co-ordinate, z is the axial co-ordinate, u is the velocity of the fluid in the axial direction relative to the duct wall, R is the radius of curvature, α is the component of $\underset{\sim}{\alpha}$ perpendicular to the canal and $\partial p/\partial z$ is the axial component of $\nabla p'$.

We now consider the possible sources of a pressure gradient. The first source we shall examine is the cupula. The cupula, when deflected, exerts a restoring force on the fluid. We model the cupula as a membrane with linear stiffness $K = \Delta p/\Delta V$, where Δp is the pressure difference across the cupula and ΔV is its volumetric displacement. If the angle subtended by the membranous duct is denoted by β, the pressure gradient in the duct produced by the cupula is

$$\frac{\partial p}{\partial z} = K\Delta V/\beta R. \tag{12}$$

Now

$$\Delta V = 2\pi \int_0^t \int_0^a u(r,t) r \, dr \, dt, \tag{13}$$

in which a is the radius of the duct.

A pressure gradient in the duct is also produced by the presence of the utricle. The size of this contribution may be assessed in the following manner. A cylinder of length ℓ being accelerated through inertial space at a rate $\ddot{s}$ will have a pressure difference between its ends of

$$\Delta p = \rho \ddot{s} \ell. \tag{14}$$

If the utricle had closed ends, the pressure difference due to rotational motion only could be approximated by

$$\Delta p \simeq \rho \gamma R^2 \alpha, \tag{15}$$

where γ is the angle subtended by the utricle. Van Buskirk (1977) has shown that the small acceleration of the fluid in the utricle relative to the walls may be ignored. He has also shown that the pressure drop in the fluid as it moves from the larger utricle into the narrow duct is negligible. Therefore (15) is an accurate approximation and the pressure gradient in the duct due to the presence of the utricle and ampulla is

$$\frac{\partial p}{\partial z} = (\gamma/\beta) \rho R \alpha. \tag{16}$$

Again, of course, we have ignored the pressure gradient associated with $\underset{\sim}{a}_c$ and $\underset{\sim}{b}$ (see (8) *et seq.*). Combining this result with the pressure gradient due to the cupula we have

$$\frac{\partial p}{\partial z} = \frac{\gamma}{\beta} \rho R \alpha + \frac{K \Delta V}{\beta R}. \tag{17}$$

Introducing (13) into (17) and substituting the resulting expression into (11), we obtain the governing equation for the fluid flow in the semicircular canals:

$$\frac{\partial u}{\partial t} + \left(1 + \frac{\gamma}{\beta}\right) R\alpha = - \frac{2\pi K}{\rho \beta R} \int_0^t \int_0^a u(r,t) r dr dt + \frac{\nu}{r} \frac{\partial}{\partial r}\left(r \frac{\partial u}{\partial r}\right) \tag{18}$$

Equation (18) is non-dimensionalized by substituting the following variables:

$$r' = r/a, \quad t' = t\nu/a^2, \quad u' = u/R\Omega,$$

where Ω is some characteristic angular velocity of the canal. In terms of these variables (18) becomes

$$\frac{\partial u'}{\partial t'} + \frac{1+\gamma/\beta}{\Omega}\alpha(t') = -\varepsilon\int_0^{t}\int_0^{1} u'r'dr'dt' + \frac{1}{r'}\left(r'\frac{\partial u'}{\partial r'}\right), \tag{19}$$

where $\varepsilon = 2K\pi a^6/\rho\beta R\nu^2$. This equation contains only one non-dimensional parameter, ε.

4. Solution

We now examine the response of the canals to two specific kinds of angular acceleration. We shall first obtain an approximate solution for the case of a step input in angular velocity, corresponding to an abrupt angular motion of the head. Since the problem is a linear one, the frequency response to sinusoidal motion of the head can be readily obtained from this first solution.

A step in angular velocity corresponds to an impulse in angular acceleration. Thus, in dimensionless form,

$$\alpha(t) = \Omega\delta(t), \tag{20}$$

where $\delta(t)$ corresponds to a unit impulse or Dirac delta function applied at $t = 0$. Substituting (20) into (19), we obtain the governing equation for the particular flow we are considering:

$$\frac{\partial u}{\partial t} + \left(1 + \frac{\gamma}{\beta}\right)\delta(t) = \frac{1}{r}\frac{\partial}{\partial r}\left(r\frac{\partial u}{\partial r}\right) - \varepsilon\int_0^{t}\int_0^{1} urdrdt. \tag{21}$$

Note that in (21), we have dropped the primes. From this point on we shall work with the non-dimensionalized form of the governing equation only.

The boundary and initial conditions for this problem are

$$u(1,t) = 0, \qquad \partial u(0,t)/\partial r = 0, \qquad u(r,0) = 0. \tag{22}$$

Before solving (21) we examine the order of magnitude of the constant ε. The dimensions of the human semicircular canal are $a = 0.15$ mm and $R = 3.2$ mm (Igarashi 1966). Studies of the physical properties of endolymph (e.g. Steer 1967) suggest that it has a viscosity and density close to those of water, i.e. $\mu = 1.0$ mPa·s and $\rho = 1000\text{kg/m}^3$. Detailed geometrical studies of the labyrinth of the cat (Fernandez & Valentinuzzi 1968) indicate that $\beta = 1.4$ and $\gamma = 0.42$. We accept these values for humans. No data are available in the literature for the cupula stiffness K of humans. For the purpose of this discussion we adopt the value $K = 3.4 \times 10^9$ Pa/m^3. (This is not entirely arbitrary, as it yields results consistent with experiment, as will be seen below.) Using these values,

$$\varepsilon = 2K\pi a^6/\rho\beta R\nu^2 = 0.017.$$

Thus ε is very small, while the normalized velocity u is of order unity. Thus the last term on the right side of (21) is very small when t is of order unity or less. However, since the term is an integral over t, it can in fact dominate when t is large. This gives the problem a *singular* character. We shall obtain a first approximation using a singular perturbation method.

4.1 *Small values of t*

The solution is assumed to take the form of a perturbation series

$$u = u^{(1)} + \varepsilon u^{(2)} + \varepsilon^2 u^{(3)} + \dots . \tag{23}$$

Substituting (23) into (21) and retaining terms of $O(\varepsilon^0)$ yields the problem

$$\frac{\partial u^{(1)}}{\partial t} + \left(1 + \frac{\gamma}{\beta}\right) \alpha(t) = \frac{1}{r}\frac{\partial}{\partial r}\left(r\frac{\partial u^{(1)}}{\partial r}\right). \tag{24}$$

The solution satisfying the initial and boundary conditions (22) has been obtained by Van Buskirk & Grant (1973), and is

$$u^{(1)} = -2\left(1 + \frac{\gamma}{\beta}\right)\sum_{n=1}^{\infty} \frac{e^{-\lambda_n t}}{\lambda_n J_1(\lambda_n)} J_0(\lambda_n r), \tag{25}$$

where J_0 is the zeroth-order Bessel function of the first kind, λ_n is the n.th root and J_1 is the first-order Bessel function of the first kind. We are interested in the displacement of the cupula. An appropriate non-dimensional measure of that displacement is the volume-flow-rate integral ϕ, given by

$$\phi = \int_0^t\int_0^1 ur\,dr\,dt \tag{26}$$

(note that $\Delta V = 2R\Omega a^4\phi/\nu$). The first approximation to ϕ (for small t) is

$$\phi^{(1)} = \int_0^t\int_0^1 u^{(1)}r\,dr\,dt = -2\left(1 + \frac{\gamma}{\beta}\right)\sum_{n=1}^{\infty} \frac{1 - e^{-\lambda_n^2 t}}{\lambda_n^4}. \tag{27}$$

4.2 *Large times, small* ε

Let $\tilde{t} = \varepsilon t$ be the stretched independent variable. When this is substituted into (21) we find that

$$\varepsilon\frac{\partial u}{\partial \tilde{t}} + \left(1 + \frac{\gamma}{\beta}\right) \delta(\tilde{t}/\varepsilon) = \frac{1}{r}\frac{\partial}{\partial r}\left(r\frac{\partial u}{\partial r}\right) - \int_0^{\tilde{t}}\int_0^1 ur\,dr\,d\tilde{t}. \tag{28}$$

The integral on the right-hand side of (28) can be written as

$$\int_0^{\tau(\varepsilon)}\int_0^1 ur\,dr\,d\tilde{t} + \int_{\tau(\varepsilon)}^{\tilde{t}}\int_0^1 ur\,dr\,d\tilde{t} \tag{29}$$

where we assume $\varepsilon/\lambda_1^2 \ll \tau(\varepsilon) \ll 1$. If ε is sufficiently small, then, according to (27), for all pratical purposes

$$\int_0^{\tau(\varepsilon)}\int_0^1 ur\,dr\,d\tilde{t} \simeq -2\left(1 + \frac{\gamma}{\beta}\right) \varepsilon\sum_{n=1}^{\infty} \frac{1}{\lambda_n^4} \tag{30}$$

and the first approximation to (28) for $t \gg \tau(\varepsilon)$ is

$$\frac{1}{r}\frac{\partial}{\partial r}\left(r\frac{\partial u^{(1)}}{\partial r}\right) = \int_{\tau(\varepsilon)}^{\tilde{t}}\int_0^1 u^{(1)}r\,dr\,d\tilde{t} - 2\left(1 + \frac{\gamma}{\beta}\right)\varepsilon\sum_{n=1}^{\infty} \frac{1}{\lambda_n^4} \tag{31}$$

Equation (31) is easily solved by using the Laplace-transform method, the result being

$$u^{(1)} = \frac{\varepsilon}{4}\sum_{n=1}^{\infty} \frac{2(1 + \gamma/\beta)}{\lambda_n^4} (1 - r^2)e^{-\frac{t\varepsilon}{16}}. \tag{32}$$

Then

$$\phi^{(1)} = \int_0^{t}\int_0^{1} u^{(1)} r\, dr\, d\tilde{t} = \left(e^{-\tau(\varepsilon)} - e^{-\frac{t\varepsilon}{16}}\right)\sum_{n=1}^{\infty}\frac{2(1+\gamma/\beta)}{\lambda_n^4}, \tag{33}$$

and since $\tau(\varepsilon) \ll 1$, we have

$$\phi^{(1)} \simeq \left(1 - e^{-\frac{t\varepsilon}{16}}\right)\sum_{n=1}^{\infty}\frac{2(1+\gamma/\beta)}{\lambda_n^4}. \tag{34}$$

Equations (27) and (34) can now be combined to yield the following *uniformly valid* first approximate solution for ϕ:

$$\phi^{(1)} = \sum_{n=1}^{\infty}\frac{2(1+\gamma/\beta)}{\lambda_n^4}\left[e^{-\lambda_n^2 t} - e^{-\frac{t\varepsilon}{16}}\right] \tag{35}$$

4.3 *Transfer function*

The transfer function relating the non-dimensional volumetric displacement ϕ to the angular acceleration α is obtained by finding the Laplace transform of ϕ and dividing by the Laplace transform of α. The first approximation is

$$\frac{\phi^{(1)}}{\alpha}(s) \simeq \sum_{n=1}^{\infty}\frac{2(1+\gamma/\beta)}{\lambda_n^4}\left(\frac{\frac{\varepsilon}{16} - \lambda_n^2}{\left(s + \frac{\varepsilon}{16}\right)\left(s + \lambda_n^2\right)}\right) \tag{36}$$

which may in turn be approximated to very high accuracy by the simpler expression

$$\frac{\phi^{(1)}}{\alpha}(s) \simeq \frac{-2(1+\gamma/\beta)/\lambda_1^2}{\left(s + \frac{\varepsilon}{16}\right)\left(s + \lambda_1^2\right)}. \tag{37}$$

This may be put into a form

$$\frac{\phi}{\alpha}(s) \simeq \frac{-2(1+\gamma/\beta)/\lambda_1^2}{(s + \tau_1^{-1})(s + \tau_2^{-1})}, \tag{38}$$

where $\tau_1 = 16/\varepsilon$, the non-dimensional long time constant, and $\tau_2 = 1/\lambda_1^2$, the non-dimensional short time constant. Transforming back to the physical time domain and expressing the transfer function in terms of θ, the mean angular displacement of the endolymph, and ω, the angular velocity of the head, we obtain

$$\frac{\theta}{\omega}(s) = \frac{-4(1+\gamma/\beta)s/\lambda_1^2}{(s + T_1^{-1})(s + T_2^{-1})}, \tag{39}$$

where $T_1 = 8\mu\beta R/K\pi a^4$ and $T_2 = a^2/\lambda_1^2\nu$.

4.4 *Frequency response*

The usefulness of a formulation such as (39) is seen when we display the frequency response of the system in terms of Bode plots. Figure 3(a) is a plot of the logarithm of the ratio of the amplitudes of θ and ω as a function of the frequency. The phase lag between θ and ω is plotted in Figure 3(b). It is clear from these plots that the semicircular canal functions as an angular velocity meter over the range $2/T_1$ and $1/2T_2$, which includes the range of physiological activity.

The important points on Figure 3 are the gain $G = 4(1 + \gamma/\beta)a^2/\lambda_1^4\nu$, the lower cut-off frequency T_1^{-1}, the upper cut-off frequency T_2^{-1}, and the natural frequency

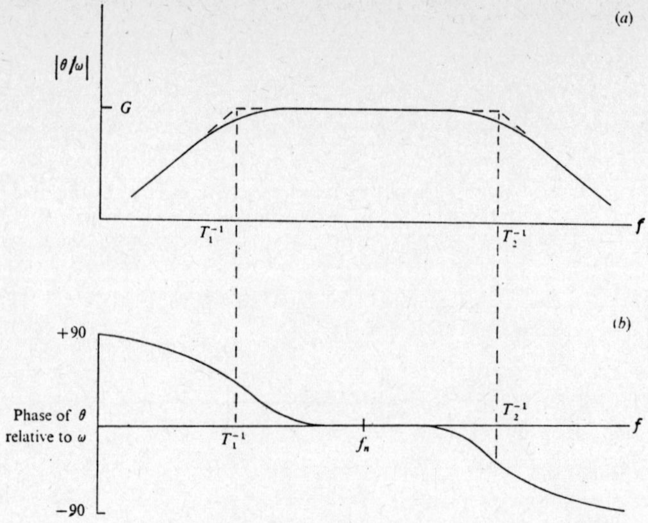

FIGURE 3. Bode plots of the gain and phase lag between the displacement of the endolymph θ and the angular velocity ω as a function of the frequency of f of sinusoidal oscillation.

$f_n = (T_1 T_2)^{-\frac{1}{2}}$. In the physical domain

$$G = 3.5 \text{ ms}, \quad T_1^{-1} = 0.048 \text{ rad/s} = 7.6 \text{ mHz}, \quad T_2^{-1} = 260 \text{ rad/s} = 41 \text{ Hz},$$

$$f_n = 3.5 \text{ rad/s} = 0.56 \text{ Hz}.$$

4.5 *Impulse response*

The impulse response of the system given by (39) is

$$\theta \simeq \frac{4(1 + \gamma/\beta)\Omega a^2}{\lambda_1^4 \nu}[e^{-t/T_2} - e^{-t/T_1}], \tag{40}$$

where Ω is the magnitude of the step in angular velocity. The initial displacement of the cupula is governed by the 'short' time constant $T_2 = 3.9$ ms and is shown in Figure 4. The maximum displacement of the endolymph is $3.5 \times 10^{-3}\Omega$ rad, where Ω is given in rad/s. The return phase of the cupula is governed by the 'long' time constant T_1 and is shown in Figure 5. $T_1 = 21$ s, which is approximately what has been observed in experimental studies.

It is interesting to note the sensation associated with such a stimulation. The time constant T_2 is too short to be 'felt' and the sensation is of an instantaneous onset of angular velocity. But then, even though the angular velocity remains at a constant level, the sensation of angular velocity decays exponentially. Obviously, one's sensation of angular velocity is linked to the displacement of the cupula.

5. Experimental observations

Several researchers have attempted to measure experimentally the fluid dynamic response of the semicircular canals. The first of these were probably van Egmond *et al.* (1949). Using subjective sensation as an indicator of canal response they measured the long time constant T_1 and the natural frequency f_n for humans. They found $T_1 \simeq 10$ s and $f_n = 0.16$ Hz. Later researchers used a more objective measure of semicircular-canal response, namely nystagmus, a characteristic movement of the eyes associated with semicircular-canal stimulation. Malcolm (1968), using nystagmus and accounting for adaptation of the central nervous system, found a mean value

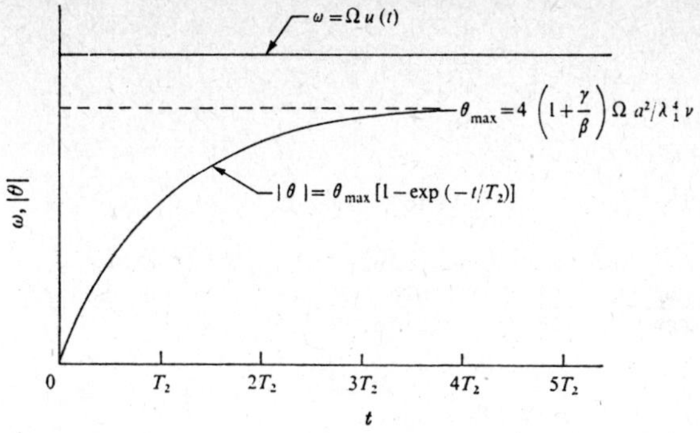

FIGURE 4. The short-term response of the endolymph to a step input in angular velocity of magnitude Ω.

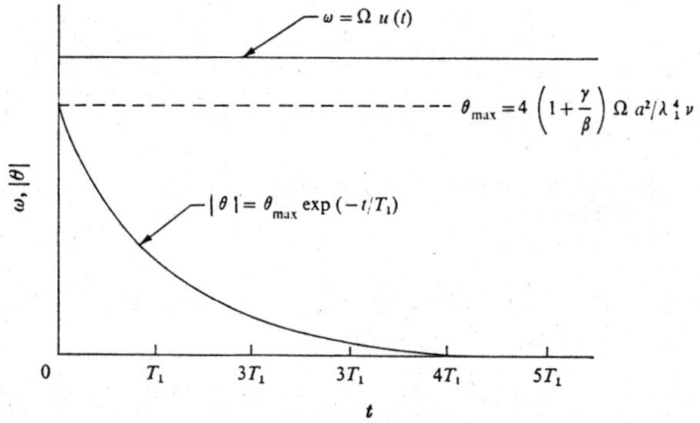

FIGURE 5. The long-term response of the endolymph to a step input in angular velocity of magnitude Ω. The time scale in this figure is much longer than that in figure 4.

of T_1 = 21 s for eight subjects. Niven & Hixson (1961), using sinusoidal oscillation as a stimulus and nystagmus as a measure of canal response, found a mean value of f_n = 0.21 Hz for six subjects.

By choosing an appropriate value for K, we have matched our theoretical time constant T_1 to that observed by Malcolm (1968). We should note in all candor, however, that our predicted value of f_n is more than twice that observed experimentally. Of course, neural processing could account for that difference, but a definitive answer will have to await further study.

6. Summary and conclusions

In this chapter we have examined the fluid dynamics of a single semicircular canal. We have shown that it responds like a heavily damped, second-order system. An examination of the frequency response shows that its function is that of an angular-velocity meter. While the canals work quite well as angular-velocity meters for ordinary motions, we see from an examination of the impulse response that an unphysiological motion can lead to the generation of erroneous signals. It has been suggested in the literature that erroneous signals to the central nervous system are the cause of spatial disorientation and motion sickness (Johnson & Jongkees 1974).

REFERENCES

EGMOND, A. A. J. VAN, GROEN, J. J. & JONGKEES, L. B. W. 1949 The mechanics of the semicircular canal. J. Physiol. 110,1.

IGARASHI, M. 1966 Dimensional study of the vestibular end of organ apparatus. In 2nd Symp. on Role of Vestibular Organs in Space Exploration. Washington, D.C.: U.S. Government Printing Office, N.A.S.A.

JOHNSON, W. J. & JONGKEES, L. B. W. 1974 Motion sickness. In Handbook of Sensory Physiology, vol. VI/2. Springer.

MALCOLM, R. 1968 A quantitative study of vestibular adaptation in humans. In 4th Symp. on Role of Vestibular Organs in Space Exploration. Washington, D.C.: U.S. Government Printing Office, N.A.S.A. SP-187

NIVEN, J. I. & HIXSON, W. C. 1961 Frequency response of the human semicircular canal. I. Steady-state ocular nystagmus response to high-level sinusoidal angular ratations. Rep. U.S. Naval School Aviation Med., Pensacola, Florida, no. NSAM-58 (N.A.S.A. Order no. R-1).

STEER, R. W. 1967 The influence of angular and linear acceleration and thermal stimulation of the human semicircular canal. Sc.D. thesis, Massachusetts Institute of Technology.

VAN BUSKIRK, W. C. 1977 The effects of the utricle on flow in the semicircular canals. Ann. Biomed. Engng 5,1.

VAN BUSKIRK, W. C. & GRANT, J. W. 1973 Biomechanics of the semicircular canals. In 1973 Biomech. Symp. New York: A.S.M.E.

THE ACOUSTICAL INVERSE PROBLEM FOR THE COCHLEA

Man Mohan Sondhi

Bell Laboratories
Murray Hill, New Jersey 07974

This article first appeared in the Journal of the Acoustical Society of America, Vol 69, No. 2, February 1981, pp 500-504.

ABSTRACT

In this article we will show how the stiffness $K(x)$ of the basilar membrane may be estimated from a measurement of the impulse response at the stapes (i.e., from a measurement of the pressure developed just inside the stapes in response to an impulse of stapes velocity.)

1. INTRODUCTION: The Direct Problem

Since the pioneering papers of Zwislocki [1] and Peterson and Bogert [2] the most enduring model for describing fluid motion in the cochlea has been the one-dimensional linear "transmission line" model. The model derives its name from the fact that under appropriate simplifying assumptions the differential equations relating fluid pressure and velocity are formally identical to those relating voltage and current in a (nonuniform) electrical transmission line.

The main assumption on which the model rests is that the fluid motion in each of the scalas (see Figure 1) is constant across any plane perpendicular to the basilar membrane (BM). This assumption is, of course, inconsistent with motion of the basilar membrane. However, at least for sinusoidal excitation, the effect of this inconsistency can be shown to be significant only near the place of resonance for that frequency.

The other usual simplifying assumptions are that the fluid filling the chambers is incompressible and inviscid, that the cross-section of the scalas is constant along the length and that the BM acts as a locally reacting partition. This last statement means that the velocity of a point on the membrane depends only on the pressure across it at that point.

Under these assumptions the differential equations of fluid motion can be written down essentially by inspection of Figure 1. Thus, let the fluid density be ρ and let $p(x,t)$ and $u(x,t)$ be the pressure and particle velocity in the scala vestibuli at time t and a distance x from the stapes. Then Newton's second law of motion gives

$$\frac{\partial p}{\partial x} = -\rho \frac{\partial u}{\partial t} \tag{1}$$

Also, since the fluid is incompressible, a change in the volume flow must be compensated by an appropriate displacement $\xi(x,t)$ of the BM. This balance is expressed by the equation

$$A\frac{\partial u}{\partial x} = W\frac{\partial \xi}{\partial t} \tag{2}$$

where A and W are the (constant) scala cross-section and BM width respectively. Note that ξ is taken to be positive upwards.

Finally, the displacement of the BM is related to the pressure across it. Assuming the BM to be linear, with mass $m(x)$, stiffness $K(x)$, and damping $R(x)$ per unit length, we have

$$m(x)\frac{\partial^2 \xi}{\partial t^2} + R(x)\frac{\partial \xi}{\partial t} + K(x)\xi = -2p. \tag{3}$$

Here we have assumed, from symmetry, that the pressure across the membrane is twice the pressure in the scala vestibuli.

The transmission line equations are obtained upon elimination of ξ from (2) and (3). The way this is usually done is in the Laplace transform (frequency) domain. Thus with quiescent initial conditions the Laplace transforms of equations (1)-(3) are:

$$\frac{\partial P}{\partial x} = -\rho s U \qquad (1)'$$

$$A\frac{\partial U}{\partial x} = Ws\,\Xi \qquad (2)'$$

$$(ms^2+Rs+K)\Xi = -2P. \qquad (3)'$$

Here $U(x,s)$, $\Xi(x,s)$ and $P(x,s)$ are, respectively the Laplace transforms of u,ξ and p. Substituting (3)' into (2)' gives

$$\frac{\partial U}{\partial x} = -\frac{2W}{A}\frac{1}{Z(x,s)}P, \tag{4}$$

where $Z(x,s) = (ms+R+\frac{K}{s})$ is the impedance per unit length of the basilar membrane. Equations (1)' and (4) are the equations of a transmission line.

For reasons that we will discuss later, we will be concerned in this paper with transient motion of the fluid (and BM) in response to impulsive excitation of the stapes. Therefore we are interested in the time domain version of (4). Defining

$$y(t) = \mathbf{L}^{-1}[\frac{W}{AZ(x,s)}] \tag{5}$$

equation (4) is seen to be

$$\frac{\partial u}{\partial x} = -2y(t) * p \tag{6}$$

where * indicates convolution (in time).

The models described in [1] and [2] are essentially equivalent to that summarized by the pair of equations (1)' and (4) or, equivalently by (1) and (6). This model is linear in view of (3) and (3)'. And because of the assumption of plane phase fronts, it is a "long wave" model. This means that for sinusoidal excitations the model is accurate only whenever the wavelength is long compared to the cross dimension ($\frac{A}{W}$). Over the years this model has been extended and modified to include the effects of BM nonlinearities, two-dimensional fluid motion , and short wavelengths. However we will not be concerned with any of these modified versions here, because at present we do not know how to modify the analysis of the next section so as to be applicable to these cases.

In order to use the above model one must specify the cochlear parameters (ρ, $\frac{A}{W}$, $m(x)$, $K(x)$, $R(x)$) as well as the boundary condition at the helicotrema. Once this has been done, the response (e.g., $\xi(x,t)$ *or* $p(x,t)$) to a velocity input at the stapes can be computed by well known methods. For example, for a sinusoidal input velocity $U_0 \sin \omega t$ at the stapes, equations (1)' and (4) are just ordinary differential equations with s fixed at $s=j\omega$. These equations are to be solved for $P(x,j\omega)$, $U(x,j\omega)$ with the given boundary condition at the helicotrema and the boundary condition $U(0,j\omega) = U_0 \sin \omega t$ at the stapes. Once $P(x,j\omega)$ is known, equation (3)' gives $\Xi(x,j\omega)$.

We will refer to this problem of finding solutions to equations (1)' and (4) or (1) and (6) as the *direct* problem for the cochlea. This is to contrast it with the *inverse* problem to be discussed in the next section. The important first step in solving the direct problem is, of course, the specification of the cochlear parameters and the boundary condition at the helicotrema. Unfortunately, this is no easy task; and it is fair to say that all published solutions to the direct problem are based on only *qualitatively* reliable estimates of these quantities.

About the only actual measurements that are extensively quoted are those of von Bekesy [3]. (The solutions of [4] are based on these measurements; see also [5].) These measurements were static measurements and were made on cochleas excised from cadavers. While spatial dimensions might be estimated accurately this way, it is highly likely that the effective values of the functions K, R, m would be rather different in a live cochlea under normal operating conditions. (Recall, for instance, that modern *in vivo* measurements of BM motion have rather significantly changed our quantitative notions regarding cochlear mechanics.) As for the boundary condition at the helicotrema, we know of no direct measurement. This boundary condition has not received much consideration because the pattern of vibration of the BM at intermediate and high frequencies is little affected by it.

Other, indirect methods have been proposed to estimate K, R and m. Thus by making reasonable hypotheses about the structure and elastic properties of the BM it is possible to compute $K(x)$ and $m(x)$ [6,7]. $R(x)$ may be similarly computed from models [8]. Also, one may assume that these functions depend on only a few parameters (e.g., $K(x)=K_0 e^{-\alpha x}$ with K_0 and α as parameters). Then it is possible [9] to find parameter values which give a good match to measurements of say BM velocity. However, invariably the measurements available are inadequate to accurately estimate the unknowns.

In the next section we will show that under certain reasonable assumptions the stiffness function of the BM may be estimated unambiguously from direct measurements. Further these measurements can be carried out *in vivo.* in Section 3 we will discuss various strong and weak points of our proposal, and will also mention possibilities of improving the estimate of the stiffness when some of the assumptions are relaxed.

2. THE INVERSE PROBLEM

Suppose we make the assumption that equations (1)-(3) adequately describe the dynamics of the cochlea. Is it then possible to infer the cochlear parameters from acoustical measurements made at accessible points? Problems of this type have attracted considerable attention in recent years in various branches of physics, and are called inverse problems. (In mathematical terms, the problem is to construct a set of differential equations from measured boundary behavior of their solutions, as opposed to finding solutions to given differential equations.)

Let us assume that the stapes end of the cochlea is all that is accessible for measurement. Let $H(t)$ be the pressure developed just inside the stapes in response to an impulse of velocity imparted to the stapes. Then in view of the assumed linearity, it is clear that a measurement of the impulse response contains all the information that acoustical measurements at the stapes can provide. This is because the response to any other input can be obtained by convolving it with $H(t)$. Thus the question is: Can a measurement of the impulse response at the stapes allow us to estimate the cochlear parameters? To this general question the answer has to be negative. Clearly it is unreasonable to expect to recover three presumably unrelated funcions K, m and R from the single function $H(t)$. Fortunately, however, the impulse response at the stapes is not much affected by the values of m and R. The reason for this happy circumstance is as follows: For a sinusoidal input at radian frequency ω, resonance occurs at a point on the BM approximately where $\omega^2=(K/m)$. Now BM stiffness *decreases* (by a factor of perhaps a hundred thousand) from the stapes towards the helicotrema; and the effective mass (BM + fluid loading) stays more or less constant. The point of resonance thus approaches the stapes end only at the highest frequency of interest. Therefore the impedance of the BM at the stapes end is predominantly stiffness controlled. Computations on models with typical parameter values confirms this expectation.

Therefore in this introductory paper on the subject, we will assume that as far as the measurement of $H(t)$ is concerned, m and R may be assumed to be zero. We will discuss some possible relaxations of this assumption in the next section. Once m and R are neglected, equations (3) and (5) simplify considerably, and the transmission line equations reduce to:

$$\frac{\partial p}{\partial x} = -\rho \frac{\partial u}{\partial t} \tag{1}$$

and

$$\frac{\partial u}{\partial x} = -\frac{2W}{A}\frac{1}{K(x)}\frac{\partial p}{\partial t}. \tag{7}$$

It turns out that the inverse problem can be solved for a transmission line governed by this set of equations. That is, $K(x)$ can be estimated from a measurement of the impulse response. To show this we will transform these equations to a form in which we have presented a solution of the inverse problem in an earlier paper [10]. (The physical problem discussed in that paper is that of determining the shape of the human vocal tract from acoustical measurements. Although that appears to be totally unrelated to the present problem, the two problems have a very strong connection from a mathematical point of view.)

The transformation that we require may be derived in the following manner.

(i) Write

$$K(x) = K_0\, k(x) \tag{8}$$

where K_0 is the stiffness of the BM per unit length at the stapes end, and $k(x)$ is a dimensionless function with $k(0)=1$.

(ii) Define the parameter

$$c = \sqrt{\frac{AK_0}{2W\rho}}. \tag{9}$$

It can be verified that c has the dimensions of a velocity. In terms of c, K_0 and $k(x)$ equation (7) becomes

$$\frac{\partial u}{\partial x} = -\frac{1}{\rho c^2}\frac{1}{k(x)}\frac{\partial p}{\partial t}. \tag{7'}$$

(iii) Finally define the change of variable x (distance along the BM) by the differential relation

$$d\hat{x} = \frac{dx}{\sqrt{k(x)}}. \tag{10}$$

with $\hat{x} = 0$ when $x=0$. In terms of this new variable let

$$\hat{p}(\hat{x},t) = p(x,t) \tag{11a}$$

$$\hat{u}(\hat{x},t)| = u(x,t) \tag{11b}$$

$$g(\hat{x}) = \frac{1}{\sqrt{k(x)}} \tag{11c}$$

It is worth noting that at the stapes ($x=\hat{x}=0$), $\hat{p}$ and $\hat{u}$ are identical to p and u respectively. With the definitions given in equations (11a)-(11c) the transmission line equations (1) and (7)' become

$$g(\hat{x})\frac{\partial \hat{p}}{\partial \hat{x}} = -\rho\frac{\partial \hat{u}}{\partial t} \tag{12a}$$

$$g(\hat{x})\frac{\partial \hat{p}}{\partial t} = -\rho c^2\frac{\partial \hat{u}}{\partial \hat{x}}. \tag{12b}$$

Equations (12) are precisely the equations relating pressure and volume velocity for a variable area tube of cross-sectional area $g(\hat{x})$, as discussed in [10]. In order to keep the presentation simple and as nonmathematical as possible, we will only quote the result from that paper, and refer the interested reader to it for details.

As shown in [10] the impulse response for a set of equations (12) has the form

$$H(t) = \rho c[\delta(t)+h(t)] \tag{13}$$

where $\delta(t)$ is the Dirac unit impulse, and $h(t)$ is continuous provided the function $g(\hat{x})$ is finite, differentiable , and greater than some strictly positive number. (We will assume that $g(\hat{x})$ satisfies these constraints.) Further, if one solves the integral equation

$$f(T,t)+\frac{1}{2}\int_{-T}^{T} h(|t-\tau|)f(T,\tau)d\tau = 1 \quad |t| \leq T \tag{14}$$

for the function $f(T,t)$ then

$$f^2(T,T) = g(cT). \tag{15}$$

Equation (11c) then gives $\sqrt{k(x)}$ for the value of x corresponding to the value $\hat{x}=cT$. Thus by solving equation (15) for various values of T, the function $\sqrt{k(x)}$ can be determined.

In the Appendix we illustrate the application of the analysis given above by deriving $k(x)$ for one impulse response for which the derivation can be carried out analytically. The impulse response chosen is

$$H(t) = \rho c[\delta(t)-ae^{-at}]. \tag{16}$$

Besides the fact that one can analytically derive $k(x)$ corresponding to it, the impulse response of eq. (16) has another interesting property. It turns out that if the solution of equations (1) and (7) is approximated by a forward travelling WKB solution (see for example [11]) then $H(t)$ has the form of equation (16) with $a=\frac{c}{4}k'(o)$. To that approximation, therefore, $H(t)$ depends only on the initial slope of $k(x)$. As we shall point out in the next section, this approximation is valid only in the mid-frequency range. The low frequency components of the impulse response are strongly affected by the stiffness function.

In general of course the integral equation (14) must be solved numerically and then the differential equation (10) must be integrated numerically. However, the solution of (14) can be obtained by readily available matrix inversion routines, and the integration of (10) likewise requires fairly straightforward numerical quadrature. In Figure (2) we show what the numerically computed impulse response looks like for the case when $k(x)$ is exponentially decreasing; that is the usually assumed functional form for the stiffness.

3. DISCUSSION AND CRITIQUE

[i] The boundary condition at the helicotrema does not appear in the solution of the inverse problem given in the previous section. This is the main motivation for carrying out the analysis in the time domain. It is intuitively reasonable (and proven rigorously in [10]) that $H(t)$ should be independent of the boundary condition at the helicotrema for a time interval less than twice the time it takes an impulsive excitation to travel down the BM from the stapes to the helicotrema. However, this is precisely the time interval for which $H(t)$ is needed in order to completely recover $k(x)$.

[ii] Note that the Fourier transform of $H(t)$ is what is called the cochlear input impedance. This quantity has been calculated from various models, but direct measurements have been few. It can be infered from measurement of intra cochlear pressure [12] vs. pressure at the tympanic membrane, provided the middle ear transfer function is also measured. It has also been measured directly [13,14]. It is true that if the frequency domain measurements are available over the infinite frequency axis, then $H(t)$ can be obtained by a Fourier transformation. However, frequency domain measurements are not the most suitable form of data for the present purpose because the boundary condition at the helicotrema is not known. This boundary condition can be quite important at low frequencies. Also important at low frequencies is the stiffness of the round window membrane. This should preferably be opened for impulse response measurements. It is to be hoped that the techniques used for frequency domain measurements can be adapted for time domain measurements.

[iii] Besides the measurement itself there is one other difficulty to be surmounted before the inverse problem can be solved in practice. That is the problem of estimating the parameter c which is unknown *a priori*.

In theory, the strength of the δ—function in equation (13) gives ρc and therefore c (since ρ is known). With c known $\rho c^2 k(x)$ may be determined as described in the previous section. This gives the stiffness function multiplied by a constant scale factor. However, it is scaled in just the manner in which it appears in the original equations.

Unfortunately the situation is not as simple as this in practice. The input velocity pulse cannot obviously be a true impulse but some band limited function. In fact, since our assumptions are not valid at high frequencies, we are forced to bandlimit the input pulse. Thus the impulse response $H(t)$ must be determined by deconvolving the pressure due to such an input. That is, by finding the function which when convolved with the input velocity yields the measured pressure. This deconvolution cannot be done accurately without a knowledge of the strength of the δ—function part of $H(t)$. A way around this dilemma is to measure the BM displacement along with the pressure at a point just inside the stapes. This provides an estimate of K_0. It remains only to estimate the effective scala height $\frac{A}{W}$

which is an easier task and, as mentioned in Section 1, need not be done *in vivo*.

Another alternative is to estimate the strength of the δ-function from the asymptotic value of the magnitude of the input impedance at high frequencies.

[iv] Finally we might mention that the estimate of $K(x)$ may be improved by taking approximate account of the quantities neglected in the analysis presented here. Thus for example the effect of viscosity can be included by adding a term $-\mu u$ on the right hand side of equation (1). If the viscosity μ is known then the effect of this perturbation can be taken into account by the method presented in [15]. The effect of including $m(x)$ and $R(x)$ can also be accurately accounted for by that method provided these functions are proportional to $K(x)$. Reference [16] shows how much more general perturbations may be accounted for in the analogous problem of the variable area tube. We surmise that these ideas too may be used for the cochlea problem; however the transformation is not obvious.

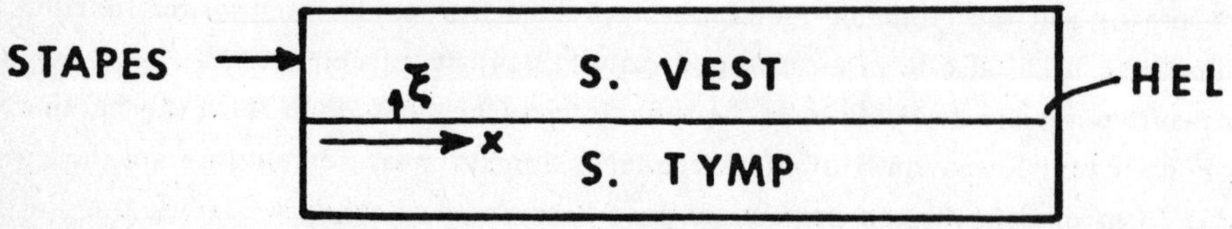

Figure 1: Schematic of the two-chamber cochlear model

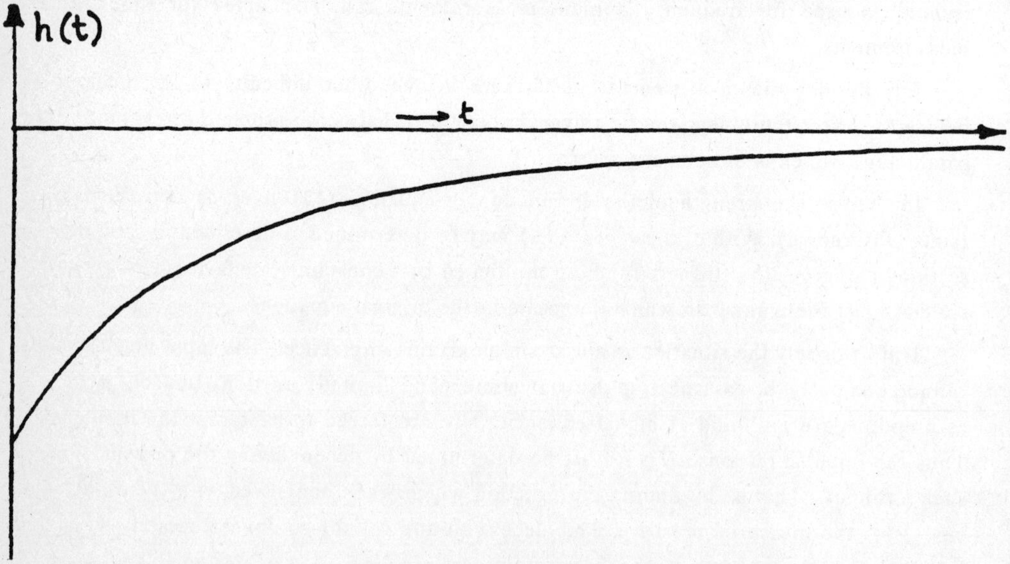

Figure 2: Illustrates the general shape to be expected for the impulse response. Curve shows the computed impulse response when k(x)=exp(-ax)

APPENDIX: AN EXAMPLE

To illustrate the steps involved in deriving $k(x)$ from $H(t)$ suppose that the measured impulse response has the form

$$H(t) = \rho c\ [\ \delta(t) - ae^{-at}\ ]. \tag{A-1}$$

This has the same general trend as the impulse response shown in Figure (2) and has the advantage that the derivation can be carried out analytically. It also has the connection with the WKB approximation mentioned in Section 2.

With $H(t)$ given by equation (A-1) some algebra shows that the solution of equation (14) of the text is

$$f(T,t) = 1+aT+\frac{a^2}{2}(T^2-t^2) \tag{A-2}$$

Then from equation (15)

$$g(cT) = f^2(T,T) = (1+aT)^2 \tag{A-3}$$

or

$$g(\hat{x}) = (1+\frac{a}{c}\hat{x})^2. \tag{A-4}$$

From equations (10) and (11c)

$$\frac{dx}{d\hat{x}} = \frac{1}{g(\hat{x})} = \frac{1}{(1+\frac{a}{c}\hat{x})^2}. \tag{A-5}$$

This is integrated easily to give

$$x = \frac{\hat{x}}{(1+\frac{a}{c}\hat{x})} \tag{A-6}$$

or

$$\hat{x} = \frac{x}{1-\frac{a}{c}x}. \tag{A-7}$$

Finally, from equations (11c) and (A-7)

$$k(x) = (1-\frac{a}{c}x)^4. \tag{A-8}$$

Of course, in order that $k(x)$ be strictly positive, as assumed in the text, this example is valid only for $x < \frac{c}{a}$.

REFERENCES

[1] Zwislocki, J., Acta Oto-Laryng., Supplement 72, (1948).
This monograph was summarized in
Zwislocki, J., "Theory of the acoustical action of the cochlea", JASA vol 22, pp 778-784 (1950).

[2] Peterson, L., and Bogert, B.P., "A dynamical theory of the cochlea", JASA vol 22, pp 369-381 (1950).

[3] von Bekesy, G., *Experiments in Hearing* McGraw Hill, N.Y., (1960), pp 473-476.

[4] Zwislocki, J., "Analysis of some audotiory characteristics", in *Handbook of Mathematical Psychology,* (R. Luce, R. Bush, and E. Galanter, eds.) vol III, Wiley, N.Y., (1965).

[5] Dallos, P., *The Auditory Periphery* Academic Press, N.Y., (1973), chapter 4.

[6] Lien, M.D., and Cox, J.R., "A mathematical model of the mechanics of the cochlea", Ph. D. dissertation, Washington U., St. Louis, Mo. (1973).

[7] Allen, J.B., "Two-dimensional cochlear fluid model: New results", JASA vol 61, pp 110-119 (1977).

[8] Allen, J.B., Talk at the Spring 1978 meeting of the Acoustical Soc. of America.

[9] Allen, J.B., and Sondhi, M.M., "Cochlear Macromechanics: Time-domain solutions", JASA vol 66, pp 123-132 (1979).

[10] Sondhi, M.M., and Gopinath, B., "Determination of Vocal-Tract Shape from Impulse Response at the Lips", JASA vol 49, pp 1867-1873 (1971)

[11] Zweig, G, Lipes, R, and Pierce, J.R., "The Cochlear Compromise", JASA vol 59, pp 975-982 (1976).

[12] Nedzelnitsky, V., "Measurements of sound pressure in the cochleae of anesthetized cats", in *Facts and Models in Hearing* E. Zwicker and E. Terhardt eds., Springer-Verlag, New York, 1974.

[13] Khanna, S.M., and Tonndorf, J., "The Vibratory Pattern of the Round Window in Cats", JASA vol 50, pp 1475-1483, (1971).

[14] Lynch, T.J., III, Nedzelnitsky, V., and Peake, W.T. "Measurement of Acoustic Input Impedance of the Cochlea in Cats", JASA vol 59, p S30, (1976).

[15] Sondhi, M.M., and Gopinath, B., "Determination of the shape of a lossy vocal tract", paper 23C10, Proc. ICA, Budapest Hungary, 1971.

[16] Resnick, J.R. "Acoustic Inverse Scattering as a means for Determining the Area Function of a Lossy Vocal Tract: Theoretical and Experimental Model Studies", Ph. D. Dissertation, Johns Hopkins University, 1979.